EDEXCEL
A LEVEL MATHS
A LEVEL

T0177693

Authors
Paul Hunt,
Steve Cavill, Rob Wagner

EXAM PRACTICE WORKBOOK

OXFORD
UNIVERSITY PRESS

Great Clarendon Street, Oxford, OX2 6DP, United Kingdom

Oxford University Press is a department of the University of Oxford.

It furthers the University's objective of excellence in research, scholarship, and education by publishing worldwide. Oxford is a registered trade mark of Oxford University Press in the UK and in certain other countries

British Library Cataloguing in Publication Data
Data available

978-0-19-841322-6

1 3 5 7 9 10 8 6 4 2

Paper used in the production of this book is a natural, recyclable product made from wood grown in sustainable forests.

The manufacturing process conforms to the environmental regulations of the country of origin.

Printed in The UK by Bell and Bain Ltd, Glasgow

Contents

About this book

This book contains three sets of write-in, mock exam papers for the Edexcel A Level Maths exam (9MA0). Full details of this exam specification can be for on the Edexcel website.

https://qualifications.pearson.com/en/qualifications/edexcel-a-levels/mathematics-2017.html

There are three papers in each exam set. Papers 1 and 2 cover Pure, and Paper 3 covers Statistics and Mechanics. All three papers are 120 minutes long and are each worth 100 marks.

The Large data set

The A Level examination will assume that you are familiar with a Large data set (LDS). In the exam, some questions will be based on the LDS and may include some extracts from it. It is Edexcel's intention that you should be taught using the LDS as this will give you a material advantage in the exam.

The data set consists of weather data samples provided by the Met Office for five UK weather stations and three overseas weather stations in the time periods May to October 1987 and May to October 2015. An Excel spreadsheet containing the LDS is available from the Edexcel website at the address given above.

Answers

The back of this book contains short answers to all the questions.
Full mark schemes for each mock paper can be found online.

https://global.oup.com/education/content/secondary/series/aqa-alevel-maths/edexcelalevelmaths-answers

Formulae

In the exam, you will be provided with a 'Formulae for A Level Mathematics' booklet that is for use in AS Level and A Level Maths qualifications. These are provided at the end of this book. The relevant A Level Maths formulae and statistical tables are provided at the end of this book.

Calculators

All papers are calculator papers. Make sure that you know how to use your calculator, particularly for statistical functions. The rules on which calculators are allowed can be found in the Joint Council for General Qualifications document 'Instructions for conducting examinations' (ICE).

| Name | | Class | |
| Signature | | Date | |

Materials

For this paper you must have:

- The booklet of formulae and statistical tables
- You may use a graphics calculator.

Question	Mark
1	
2	
3	
4	
5	
6	
7	
8	
9	
Total	

Instructions

- Use black ink or black ball-point pen.
  Pencil should be used for drawing.
- Answer **all** questions.
- You must answer each question in the space provided for that question. If you require extra space, use a supplementary answer book; do **not** use the space provided for a different question.
- Do not write outside the box around each page.
- Show all necessary working; otherwise marks for method may be lost.
- Do all rough work in this book. Cross through any work that you do not want to be marked.

Information

- The marks for questions are shown in brackets.
- The maximum mark for this paper is 100.

Advice

- Unless stated otherwise, you may quote formulae, without proof, from the booklet.
- You do not necessarily have to use all the space provided.

Answer **all** questions in the spaces provided.

1   Find $\int \ln 2x \, \mathrm{d}x$

You must show your working.   **[3 marks]**

_____

_____

_____

_____

_____

_____

_____

_____

_____

_____

_____

_____

2   The diagram shows a sector ABC with radius r and angle θ, where θ is in radians.

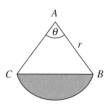

The arc length BC is P cm and the sector area ABC is Q cm²

It is given that $Q = 3P$

a   Find the length r **[3 marks]**

_____

_____

_____

_____

_____

_____

_____

_____

2 b It is also given that the triangle *ABC* is equilateral.

Find the exact value of the area of the shaded segment.

You must fully justify your working. **[5 marks]**

3 A quadrilateral *ABCD* is formed by joining the points of intersection of the lines with equations

$$y = 2x + 1$$
$$y - 2x = -10$$
$$2y = 1 - 4x$$
$$y + 2x - 6 = 0$$

a i Write down the gradients of each of the four straight lines. **[2 marks]**

ii What can you deduce about the shape of quadrilateral *ABCD*?

You must justify your answer. **[2 marks]**

3 **b** **i** Describe how you would find the coordinates of the four vertices of the quadrilateral.

Do not include any calculations at this stage. **[2 marks]**

ii Find the exact values of the coordinates of each of the four vertices of the quadrilateral.

[4 marks]

3 c   If $AB < BC$, what are lengths of sides AB and BC respectively in centimetres?   **[4 marks]**

4 a   Prove, by contradiction, that $\sqrt{2}$ is irrational.   **[6 marks]**

4 b Simplify $\dfrac{\sqrt{2\sqrt{2\sqrt{2}}}}{\sqrt{2\sqrt{2}}}$, giving your answer in the form $\sqrt[a]{2}$, where a is an integer. **[3 marks]**

c A geometric sequence has first three terms 8, b and 4.

i Find the exact value of b in the form $m\sqrt{n}$, where m and n are integers. **[3 marks]**

4 c ii Find the sum to infinity of the geometric sequence, showing all of your working.

Give your answer in the form $f + g\sqrt{h}$, where f, g and h are integers. **[5 marks]**

5 a Find the values of k such that $kx^2 + 4x + 5 = k$ has **no** real solutions.

You must show all your working. **[5 marks]**

5 b $(x-p)$ is a factor of $3x^2 -(p+8)x-(p+18)$, where p is an integer.

i   Find the possible value(s) of p

You must show all your working.   **[6 marks]**

ii For each value of p found in part **b i**, find the other factor of the quadratic expression.   **[2 marks]**

6   a   Solve the equation $\dfrac{14}{x} - x = 5$

You must show each step of your working.   **[3 marks]**

b   Write down the equation of the line l shown in the diagram below.   **[2 marks]**

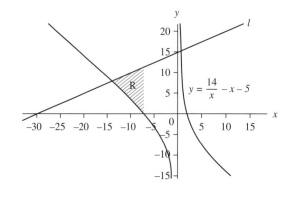

$$y = \frac{14}{x} - x - 5$$

6 c   By considering your answers to **a** and **b**, use calculus to find the area of the region labelled R.

Give your answer in the form $\dfrac{a}{b} + c\ln 2$, where a, b and c are integers.   **[11 marks]**

7 **a** Starting with the identity $\cos^2\theta + \sin^2\theta \equiv 1$, prove that $1 + \tan^2\theta \equiv \sec^2 x$ **[2 marks]**

b By using the identity from part **a**, or otherwise, solve the equation

$\sec^2\theta - \sec\theta = 1$ for $-\pi \le \theta \le 2\pi$

Give all values of θ in radians correct to three significant figures.

You must show every step of your working.   **[5 marks]**

7 c   Hence solve, for $-\pi \leq x \leq 2\pi$, the equation $\sec^2\left(\sin\frac{1}{2}x\right) - \sec\left(\sin\frac{1}{2}x\right) = 1$

Give your answers to three significant figures.   **[4 marks]**

8 a   Show that the equation $x^2 - 6 = 0$ has a root between $x = 2.4$ and $x = 2.5$ **[3 marks]**

8   b   Hence, starting with $x_0 = 2.4$, use the Newton–Raphson method **once** to find an approximate value of $\sqrt{6}$ **[3 marks]**

8   c   By defining a suitable function and then using the Newton–Raphson method (starting with $x_0 = 1.3$), find an approximation to the value of $\sqrt[7]{7}$ correct to six decimal places.   **[5 marks]**

8   d

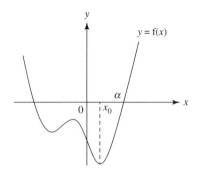

Jolene wishes to use the Newton–Raphson method to find an approximation to the positive root, α, of the curve $y = f(x)$ (shown in the diagram).

She proposes starting the procedure at the point x_0, as shown in the diagram.

Will Jolene's method prove to be successful?

Justify your answer.   **[2 marks]**

9 **a** Given that $x = \tan y$, find $\dfrac{dx}{dy}$ as a function of y (you must fully justify your working). **[1 mark]**

b Hence find $\dfrac{d}{dx}(\tan^{-1} x)$ as a function of x **[4 marks]**

End of questions

| Name | _____ | Class | _____ |
| Signature | _____ | Date | _____ |

Materials

For this paper you must have:

- The booklet of formulae and statistical tables
- You may use a graphics calculator.

Instructions

- Use black ink or black ball-point pen.
  Pencil should be used for drawing.
- Answer **all** questions.
- You must answer each question in the space provided for that question. If you require extra space, use a supplementary answer book; do **not** use the space provided for a different question.
- Do not write outside the box around each page.
- Show all necessary working; otherwise marks for method may be lost.
- Do all rough work in this book. Cross through any work that you do not want to be marked.

Question	Mark
1	
2	
3	
4	
5	
6	
7	
8	
9	
10	
11	
12	
Total	

Information

- The marks for questions are shown in brackets.
- The maximum mark for this paper is 100.

Advice

- Unless stated otherwise, you may quote formulae, without proof, from the booklet.
- You do not necessarily have to use all the space provided.

1   If $y = x \ln 5$, what is $\dfrac{dy}{dx}$?   **[1 mark]**

2   The function $f(x) = x^3$ is mapped onto the function $g(x) = 8x^3$

Express this mapping as a **single** transformation in **two** different ways.   **[3 marks]**

3 By completing the square, find the radius and centre of the circle with equation

$3x^2 + 3y^2 + 6x - 12y + 4 = 0$

You must show all your working. **[6 marks]**

4 Express $\dfrac{40-3x}{x^2(x^2-9x+20)}$ as the sum of partial fractions. **[8 marks]**

5   The graph of $y = f(x)$ is shown for $-2\pi \leq x \leq 2\pi$

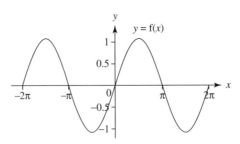

a **i** On the set of axes below, draw the graph of the gradient function, $y = f'(x)$

Label the coordinates of the axes intercepts and the turning points.   **[2 marks]**

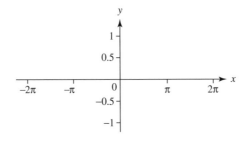

    ii Suggest a suitable equation for the graph of $y = f'(x)$ **[1 mark]**

5 b i On the set of axes below, draw the graph of the gradient of the gradient function, $y = f''(x)$

[2 marks]

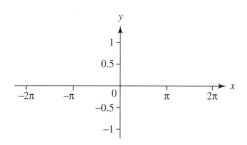

ii Suggest a suitable equation for the graph of $y = f''(x)$ [1 mark]

6   The square of the radius of a circle is plotted against its area.

This is then repeated for several other circles on the same set of axes.

The resulting graph shows a linear relationship, as seen below:

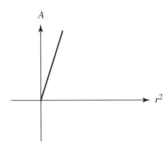

a   What is the **exact** value of the gradient of the line?   [1 mark]

_____

_____

_____

_____

b   Why is the graph only drawn in the first quadrant?   [1 mark]

_____

_____

_____

_____

_____

_____

6 c   The circle shown below has equation $x^2 + y^2 = r^2$

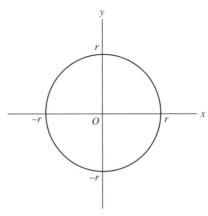

   i   By changing the subject to y and then using the substitution $x = r\sin\theta$,
           show that the area enclosed between the curve and the positive x- and y-axes is

$$r^2 \int_0^{\frac{\pi}{2}} \cos^2 \theta \; \mathrm{d}\theta$$

[6 marks]

6   c   ii   Evaluate the area enclosed between the curve and the positive x- and y-axes.   **[4 marks]**

6   d   Hence write down the area of the full circle.   **[1 mark]**

7 The series below is the sum of a geometric progression and an arithmetic progression.

$32\,768 - 700 - 16\,384 - 740 + 8192 - 780 - 4096 - 820 + ...$

The series has 40 terms in total.

a What is the 30th term of the series? [2 marks]

b What is the 37th term of the series? [2 marks]

c What is the exact value of the sum of all 40 terms of the series? [5 marks]

8   A vector **p** has magnitude 10 N and direction (bearing) $000°$

A second vector **q** has magnitude 25 N and direction $220°$

Work out the magnitude and direction of the resultant vectors of **p** and **q**

Give your answers to 1 decimal place.   **[6 marks]**

9   a   On the same set of axes, sketch the curves $f(x)=\left|e^{x}-4\right|$ and $g(x)=\left|e^{x}\right|-4$

Label each curve with its respective equation. Also, clearly indicate any points where either graph meets the coordinate axes.   **[4 marks]**

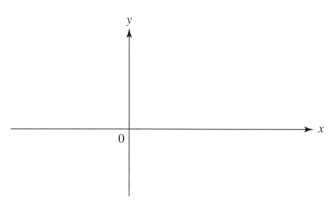

b   Hence shade the region on your graph where $y\le\left|e^{x}-4\right|$ **and** $y\ge\left|e^{x}\right|-4$ **[1 mark]**

....

....

....

....

....

10 a   Show that the distance, D, between the origin and a point (x, y) on the line $y = 4x + 3$ can be written as

$$D = \sqrt{17x^2 + 24x + 9}$$

[3 marks]

b   i   Use a method of calculus to find the exact values of x and y for which D^2 has a minimum value. You must justify that these values of x and y minimise D^2

[7 marks]

10 b ii Hence find the minimum distance between the line $y = 4x + 3$ and the origin. **[2 marks]**

11 a i $f(x) = \sin^{-1} x$ and $g(x) = \cos^{-1} x$

State the maximum domain, and corresponding range, of $f(x)$ and $g(x)$

Domain $f(x)$ _____

Range $f(x)$ _____

Domain $g(x)$ _____

Range $g(x)$ _____

[4 marks]

ii For the domains you identified in **i**, sketch the graphs of $y = f(x)$ and $y = g(x)$ on the separate axes below.

On your sketches, mark the coordinates of any points where the graphs intersect the axes.

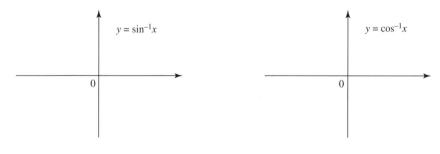

$y = \sin^{-1} x$

$y = \cos^{-1} x$

[2 marks]

11 b Using the substitution $u = \sin^{-1} x$, followed by a suitable identity, find the exact value of $\sin^{-1} x + \cos^{-1} x$ for any real x **[6 marks]**

c Prove that the value of $\cos(2\sin^{-1} x)$ is $1 - 2x^2$ **[4 marks]**

12 The diagram below shows two concentric circles, C_1 and C_2

Circle C_1 has equation $x^2 + y^2 - 2x - 4y + 1 = 0$

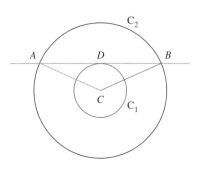

a By completing the square, find the radius of circle C_1 and the coordinates of its centre, C

[3 marks]

Circle C_1 has a horizontal tangent which meets the circle at D

b i Write down the equation of this tangent.

[1 mark]

12 b ii Given that circle C_2 has equation $x^2 + y^2 - 2x - 4y = 20$, find the exact length of the line segment AB **[5 marks]**

iii Hence find the perimeter of the segment AB

Give your answer correct to 3 significant figures. **[6 marks]**

End of questions

A Level Mathematics
Paper 3 (Set A)

Edexcel

| Name | _____ | Class | _____ |
| Signature | _____ | Date | _____ |

Candidates may use any calculator permitted by ICE regulations. Calculators must not have the facility for algebraic manipulation, differentiation and integration, or have retrievable mathematical formulae stored in them.

Question	Mark
1	
2	
3	
4	
5	
6	
7	
8	
9	
10	
11	
12	
13	
14	
15	
Total	

Instructions

- Use black ink or ball-point pen.
- If pencil is used for diagrams/sketches/graphs it must be dark (HB or B).
- **Fill in the boxes** at the top of this page with your name and class.
- Answer all the questions and ensure that your answers to parts of questions are clearly labelled.
- Answer the questions in the spaces provided
  – *there may be more space than you need.*
- You should show sufficient working to make your methods clear. Answers without working may not gain full credit.
- Inexact answers should be given to three significant figures unless otherwise stated.

Information

- A booklet 'Mathematical Formulae and Statistical Tables' is provided.
- There are 15 questions in this question paper. The total mark for this paper is 100.
- The marks for each question are shown in brackets
  – *use this as a guide as to how much time to spend on each question.*

Advice

- Read each question carefully before you start to answer it.
- Try to answer every question.
- Check your answers if you have time at the end.

Section A

Answer **all** questions in the spaces provided.

1 X is a random variable with distribution $X \sim N(10,16)$

Calculate:

a $P(X \leq 6)$ **[1 mark]**

_____

_____

_____

b $P(X \geq 16)$ **[1 mark]**

_____

_____

_____

2   *S* is a sample space for a random experiment. The probabilities for events *X* and *Y* follow this tree diagram.

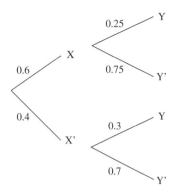

In the space provided below, draw a Venn diagram to show the same information.   **[3 marks]**

3   A six-sided dice is rolled 270 times to see if it is a fair dice. The following results are obtained.

Outcome	1	2	3	4	5	6
Frequency	41	35	46	72	31	45

Does this suggest that the dice is fair or not? Justify your answer.   **[2 marks]**

4   A sports stadium checks all spectators' bags for security purposes as they enter the stadium.

The probability that a random person entering the stadium has brought a bag is 0.3. The stadium has a capacity of 37 890 people.

a   Let the random variable X represent the number of bags which need to be checked by security at a single sporting event.

State the distribution of X, including any parameters.   **[1 mark]**

b   Explain why X can be reasonably approximated by a Normally distributed variable Y, and state the distribution of Y, including any parameters.   **[3 marks]**

4 c Use your random variable Y to show that the probability of the number of bags being checked is between 11 000 and 11 500 (inclusive) is 0.9327 to 4 dp. **[4 marks]**

_____

_____

_____

_____

_____

_____

_____

_____

_____

_____

_____

_____

_____

5 Events A, B, and C are such that:

$P(A) = 0.5$, $P(B) = 0.7$, $P(C) = 0.6$

$P(A \cap B \cap C) = 0.13$, $P(A|C) = \dfrac{1}{3}$, $P(B|A) = 0.78$, $P(B \cap C) = 0.36$

The probabilities are to be represented on the Venn diagram below.

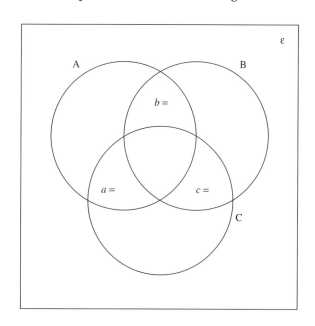

a Write the value of a on the Venn Diagram. **[1 mark]**

b Write the value of b on the Venn Diagram. **[1 mark]**

c Write the value of c on the Venn Diagram. **[1 mark]**

d Complete the Venn diagram by writing the correct probability in each blank space. **[3 marks]**

6 Rae is studying the levels of sunshine in England. She takes the daily total number of hours of sunshine recorded at weather stations in Heathrow and Leeming over a period of six months (184 days) in 2015.

She obtains the following results.

Daily total sunshine h (hours)	$0 < h \leq 0.1$	$0.1 < h \leq 0.5$	$0.5 < h \leq 2$	$2 < h \leq 3$	$3 < h \leq 5$	$5 < h \leq 7$	$7 < h \leq 9$	$9 < h \leq 12$	$12 < h \leq 15$
Heathrow	15	14	21	14	25	34	21	30	10
Leeming	18	6	15	20	34	39	19	23	10

a Is the data discrete or continuous? Give a reason for your answer. **[1 mark]**

_____

_____

_____

b Estimate the median daily number of hours of sunshine recorded at Heathrow and the median daily number of hours of sunshine recorded at Leeming. **[3 marks]**

_____

_____

_____

_____

_____

_____

c Estimate the mean daily number of hours of sunshine recorded at Heathrow and the mean daily number of hours of sunshine recorded at Leeming. **[4 marks]**

_____

_____

_____

_____

_____

_____

6 d   Plot a frequency polygon for the daily number of hours of sunshine recorded
at Leeming.

[3 marks]

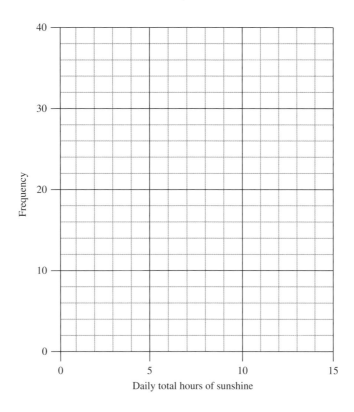

7   Henry is studying weather conditions across the world.

He looks at the daily total rainfall (in mm) and daily mean windspeed (in knots) at 8 locations worldwide on 26th June 1987

He obtains the following results.

	Cambourne	Heathrow	Hurn	Leeming	Leuchars	Beijing	Jacksonville	Perth
Daily Total Rainfall (mm)	10.7	0.3	1.6	4	0	2	2	3
Daily Mean Windspeed (knots)	10	6	8	5	5	4.5	7	1.833 333

a   For the **daily total rainfall** amounts, calculate:

  i   The median,   **[1 mark]**

  ii   The interquartile range (IQR).   **[1 mark]**

b

  i   By referring to the interquartile range (IQR), give a definition of an outlier.   **[1 mark]**

7 b ii Determine whether there are any outliers in the sample of **daily total rainfall** amounts. **[2 marks]**

_____

_____

_____

Henry drew the following scatter diagram to show the **daily total rainfall levels** against the **daily mean windspeed levels.**

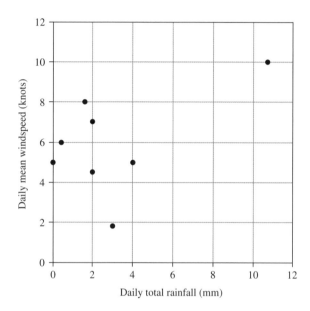

c Circle the plotted point corresponding to Beijing. **[1 mark]**

Henry calculates that the equation of the regression line of l on n is $l = 4.8 + 0.38n$

d i Draw Henry's regression line on the scatter diagram above. **[2 marks]**

ii Describe the correlation between **daily total rainfall levels** and **daily mean windspeed levels** shown by Henry's results. **[2 marks]**

_____

_____

_____

iii Use the regression line to predict the daily mean windspeed for a city whose total rainfall that day is 6 mm. **[2 marks]**

_____

_____

7 d iv Give a reason why your prediction might not be accurate. **[1 mark]**

Henry took the sample to determine if there is any correlation between the levels of rainfall and windspeed.

e i State Henry's hypotheses clearly. **[2 marks]**

ii The test statistic is 0.520 and the critical values are ± 0.754 at the 5% level.
Determine, in context, the conclusion to Henry's test. **[3 marks]**

End of section A

Section B

Answer **all** questions in the spaces provided.

8   A box of mass 20 kg rests on a horizontal table.
What is the reaction of the table on the box?   **[1 mark]**

_____

_____

9   A force of 25 N acts 1.3 m from a point *A* at an angle 72° to the horizontal line running through *A*, as shown in the diagram.

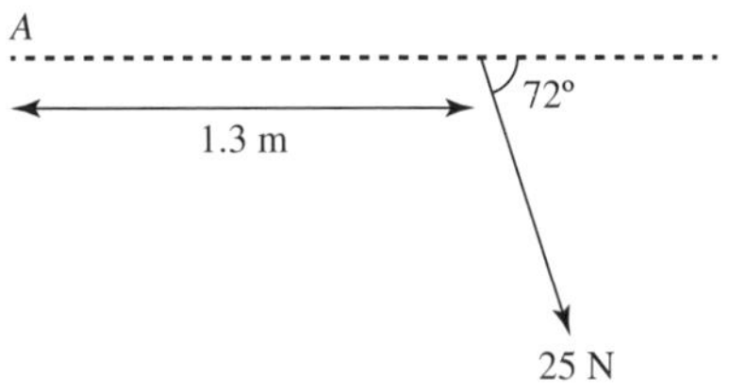

Calculate the moment of this force about *A*

You must show your working.   **[2 marks]**

_____

_____

_____

_____

_____

_____

10 a A particle accelerates uniformly from u ms^{-1} to v ms^{-1} in t seconds. The motion of the particle is shown in the velocity-time graph below.

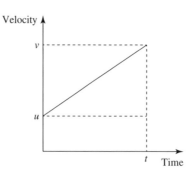

i Describe how you could use the graph to find the acceleration of the particle. **[1 mark]**

_____

ii Using your answer to **i**, show that $v = u + at$ **[3 marks]**

_____

_____

_____

_____

_____

b A car is initially moving at 50 km h^{-1}

It decelerates at a constant rate of 4 ms^{-2} until its velocity is 20 km h^{-1}

To the nearest metre, what distance does the car travel during this motion? **[3 marks]**

_____

_____

_____

_____

_____

11   A uniform ladder of length 4.8 m and mass 20 kg stands on rough, horizontal ground and rests against a smooth, vertical wall. The ladder makes an angle of 70° with the ground.

a   In which direction does the wall exert a reaction force on the ladder?

Give a reason for your answer.   **[2 marks]**

_____

_____

_____

_____

_____

_____

_____

b   Calculate the magnitude of the reaction force acting on the top of the ladder.   **[4 marks]**

_____

_____

_____

_____

_____

_____

_____

_____

_____

_____

_____

_____

12   A particle of mass 6 kg is acted on by forces of $(5\mathbf{i}+7\mathbf{j}+3\mathbf{k})$ N, $(2\mathbf{i}-\mathbf{j}+6\mathbf{k})$ N and $(\mathbf{i}-\mathbf{k})$ N

Find the magnitude of the acceleration of the particle.   **[4 marks]**

13   A firework is projected from a point on horizontal ground.
The initial velocity of the firework is 150 ms^{-1} and it is projected at an angle q to the horizontal.

There is a horizontal layer of thick cloud at a height of 0.5 km above the ground. In order for the audience to see the firework, θ must be chosen so that the path of the firework remains fully below the cloud.

a   Calculate the value of θ for which the missile *just* reaches the cloud.

You must show all your working.   **[3 marks]**

b   The technician makes a mistake and launches the firework when $\theta = 45°$

Assuming the firework does not explode, find, to the nearest tenth of a second, the time during which the firework is hidden by the cloud.

Show your working.   **[4 marks]**

14   A box of mass 5 kg is placed on a rough slope inclined at 40° to the horizontal. It is released from rest and slides down the slope.

    a   Draw a diagram to show the forces acting on the box.   **[2 marks]**

    b   Find the magnitude of the normal reaction acting on the box.   **[2 marks]**

    c   The coefficient of friction between the box and the slope is 0.3

         Find the acceleration of the box.   **[4 marks]**

15 A particle moves in a horizontal plane, where $\mathbf{i}$ and $\mathbf{j}$ are perpendicular unit vectors. At time t seconds the particle's velocity v ms^{-1} is given by $\mathbf{v} = 20\sin\left(\dfrac{\pi}{6}t\right)\mathbf{i} - 6\sqrt{t}\,\mathbf{j}$

a Find an expression for the acceleration of the particle at time t seconds. **[3 marks]**

The particle, which has mass 5 kg, moves under the action of a single force of magnitude $\mathbf{F}$ N

b i Find an expression for $\mathbf{F}$ in terms of t **[2 marks]**

ii Show that, when $t = 9$, the magnitude of $\mathbf{F}$ is 5

You should also state the direction in which $\mathbf{F}$ acts. **[4 marks]**

15 c   When $t = 9$ the particle is at the point with position vector $100\mathbf{i} + 100\mathbf{j}$

Find an expression for the position vector, $\mathbf{r}$ m, of the particle at time t **[6 marks]**

End of questions

| Name | _____ | Class | _____ |
| Signature | _____ | Date | _____ |

Materials

For this paper you must have:

- The booklet of formulae and statistical tables
- You may use a graphics calculator.

Instructions

- Use black ink or black ball-point pen.
  Pencil should be used for drawing.
- Answer **all** questions.
- You must answer each question in the space provided for that question. If you require extra space, use a supplementary answer book; do **not** use the space provided for a different question.
- Do not write outside the box around each page.
- Show all necessary working; otherwise marks for method may be lost.
- Do all rough work in this book. Cross through any work that you do not want to be marked.

Question	Mark
1	
2	
3	
4	
5	
6	
7	
8	
9	
10	
11	
Total	

Information

- The marks for questions are shown in brackets.
- The maximum mark for this paper is 100.

Advice

- Unless stated otherwise, you may quote formulae, without proof, from the booklet.
- You do not necessarily have to use all the space provided.

1   The lines $y = x$ and $y = -x$ are sketched below.

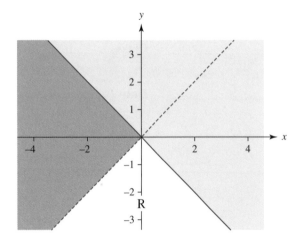

Write down two inequalities which together describe the unshaded region R.   **[2 marks]**

2 If $y = (4x-3)^4(3-5x)^2$, show that $\dfrac{dy}{dx} = 6(3-5x)(13-20x)(4x-3)^3$

You should clearly show all of your working.   **[4 marks]**

If $y = (4x-3)^4(3-5x)^2$, show that $\dfrac{dy}{dx} = 6(3-5x)(13-20x)(4x-3)^3$

You should clearly show all of your working.

3 The graph of the function $y = \dfrac{2}{x}$ is sketched below.

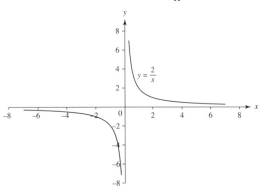

Sketch the following graphs on the axes provided below, clearly giving the coordinates of any points where your graphs intersect the coordinates axes, and the equations of any asymptotes:

a $y = \dfrac{2}{x+2}$

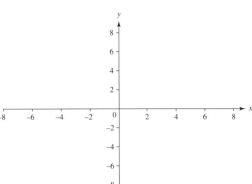

[2 marks]

b $y = -\dfrac{2}{x+2}$

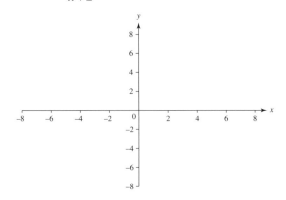

[2 marks]

3 c $y = \dfrac{2}{|x+2|} - 2$

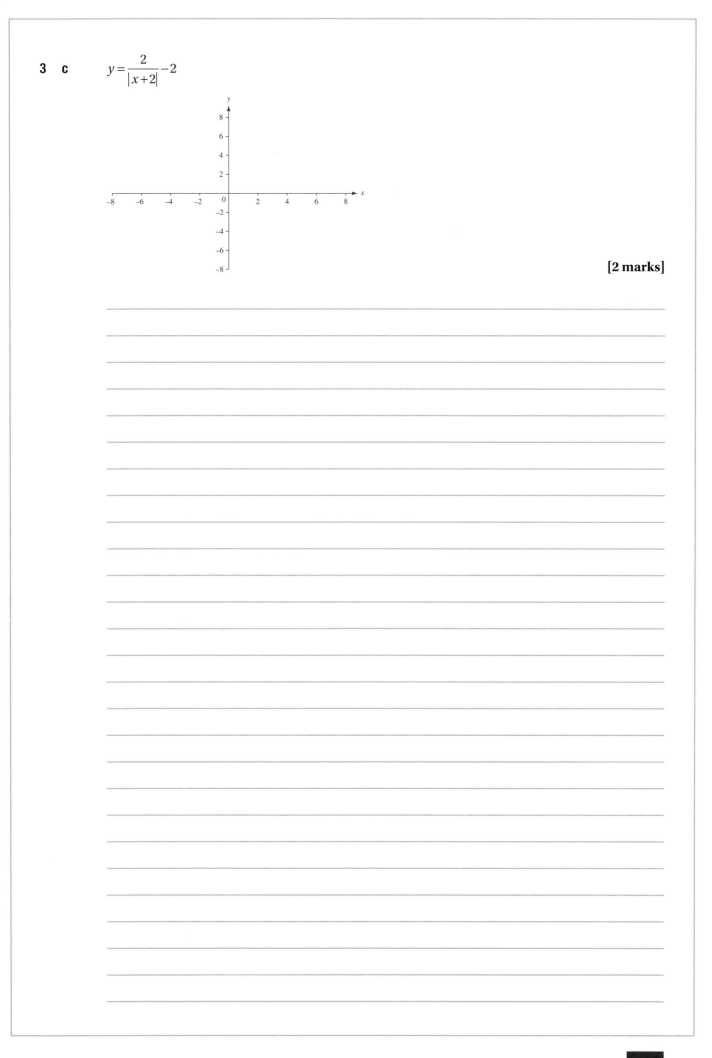

4 **a** A curve, C_1, is defined by the parametric equations $x = \dfrac{t}{t^2+1}$, $y = \dfrac{1}{t^2+1}$

Show that the Cartesian equation of the C_1 is $x^2 + y^2 - y = 0$ **[4 marks]**

b Another curve, C_2, is defined by the parametric equations $x = \tan\theta + 3$, $y\cos\theta = 4$

Find a Cartesian equation for C_2 **[4 marks]**

5   Newton's law of cooling states that the rate at which the temperature T of a body decreases over time t minutes is proportional to the difference between the temperature of the body and the temperature of its surroundings, T_0

a   Explain how this leads to the first order differential equation

$$\frac{dT}{dt} = -k(T - T_0)$$

where k is a positive constant.   **[2 marks]**

b   A freshly brewed cup of oolong tea is set to rest at an initial temperature of $80\,°C$

The surrounding room temperature is a steady $20\,°C$

The tea takes 10 minutes to cool from $80\,°C$ to $50\,°C$

i   Find the value of the constant of proportionality in the form $a\ln b$, where a and b are rational numbers.   **[8 marks]**

5 b ii What will be the temperature of the oolong tea after it has been cooling for 20 minutes?

[3 marks]

iii How long will it take for the oolong tea to cool down to 30 °C?

Give your answer to the nearest second.

[3 marks]

5 **c**   What will happen to T as $t \to \infty$?

Why does your answer seem sensible? **[2 marks]**

6 **a**   Daniel claims that:

'64 is the only number between 1 and 1000 that is both a square and a cube number.'

Find a counter-example to show that Daniel's claim is incorrect. **[1 mark]**

6 b   Fiona states that:

'Every integer between 50 and 60 (inclusive) is the sum of at most 3 triangular numbers.'

Use the method of exhaustion to prove whether or not Fiona's statement is true.   **[3 marks]**

6 c   Prove by contradiction that there is an infinite number of prime numbers.   **[7 marks]**

7 a i Find $\int \ln x \, dx$ [4 marks]

ii Hence show that $\int (\ln x)^2 \, dx = x(\ln x)^2 + 2x(1 - \ln x) + d$, where d is an arbitrary constant.

Show **all** of your working clearly. [4 marks]

7   b   Find the **exact** value of the area bounded by the curves $y = \ln x$ and $y = (\ln x)^2$, as shown shaded on the graph below.

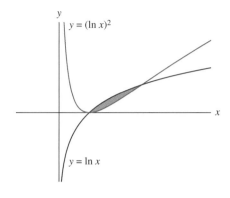

[7 marks]

8 a Express the sum $\sin\theta+\sqrt{3}\cos\theta$ in the form $R\sin(\theta+\alpha)$, where $R>0$ and $0<\theta<\dfrac{\pi}{2}$ **[3 marks]**

b Hence, find the *exact* solutions to the equation

$$\sin2\theta+\sqrt{3}(1-2\sin^2\theta)=2 \quad \text{for } -2\pi<\theta<2\pi$$

To gain full marks, you must show every step of your working. **[7 marks]**

9 **a** Find the binomial expansion of $(64 - x)^{\frac{1}{6}}$ up to and including the term in x^2 [4 marks]

b State the values of x for which the expansion in part **a** is valid.   [2 marks]

9 **c** **Hence** find an approximation to $\sqrt[6]{63}$, giving your answer to four decimal places. **[2 marks]**

10 Relative to an origin, O, the position vectors of the points A, B, C and D are

$$\overrightarrow{OA} = \mathbf{i} - \mathbf{j} + \mathbf{k}, \ \overrightarrow{OB} = 5\mathbf{i} + 2\mathbf{j} - \mathbf{k}, \ \overrightarrow{OC} = 7\mathbf{i} + 5\mathbf{j} - 3\mathbf{k} \ \text{and} \ \overrightarrow{OD} = 3\mathbf{i} + 2\mathbf{j} - \mathbf{k}$$

a Prove that the quadrilateral $ABCD$ is a parallelogram. **[3 marks]**

10 b   What are the lengths of the sides of parallelogram $ABCD$?   [2 marks]

10 c M is the midpoint of side BC

N is the point of intersection of AM and BD

Let $\overrightarrow{BN} = \mu\overrightarrow{BD}$, where μ is a constant and $\mu < 1$

Also, let $\overrightarrow{MN} = \lambda\overrightarrow{MA}$, where λ is a constant and $\lambda < 1$

i   Find a direction vector for $\overrightarrow{BN}$ in terms of **i**, **j**, **k** and μ [2 marks]

ii   Find a direction vector for $\overrightarrow{MN}$ in terms of **i**, **j**, **k** and λ [2 marks]

11 a If $f(x)=\sqrt{x}$ and $g(x)=\sqrt[6]{x}$, find the composite function $fg(x)$

Give your answer in the form $fg(x)=x^{\frac{a}{b}}$, where a and b are integers. **[2 marks]**

b If $F(x)=x+1$, $G(x)=\dfrac{1}{x+1}$ and $H(x)=\dfrac{1}{x-1}$ find the composite function $FGH(x)$

Write your answer in the form $\dfrac{ax+b}{cx+d}$, where a, b, c and d are integers. **[4 marks]**

11 c $h(x) = 2x - 1$ and $k(x) = 2x^2 - 10x - 1$

Find a function $j(x)$ such that $hj(x) = k(x)$ [3 marks]

End of questions

| Name | | Class | |
| Signature | | Date | |

Materials

For this paper you must have:

- The booklet of formulae and statistical tables
- You may use a graphics calculator.

Instructions

- Use black ink or black ball-point pen.
  Pencil should be used for drawing.
- Answer **all** questions.
- You must answer each question in the space provided for that question. If you require extra space, use a supplementary answer book; do **not** use the space provided for a different question.
- Do not write outside the box around each page.
- Show all necessary working; otherwise marks for method may be lost.
- Do all rough work in this book. Cross through any work that you do not want to be marked.

Information

- The marks for questions are shown in brackets.
- The maximum mark for this paper is 100.

Advice

- Unless stated otherwise, you may quote formulae, without proof, from the booklet.
- You do not necessarily have to use all the space provided.

Question	Mark
1	
2	
3	
4	
5	
6	
7	
8	
9	
10	
11	
12	
13	
14	
Total	

Answer **all** questions in the spaces provided.

1 What is the minimum value of $5 + \cos 3\theta$? **[1 mark]**

2 Given that $3^{2x-1} = 20$ show, by taking logs of both sides, that $x = \dfrac{\ln 60}{\ln 9}$ **[3 marks]**

3 If $\int_1^2 f(x)\,dx = 3$, what is the value of $\int_1^2 2[f(x)+1]\,dx$?

You must show all your working clearly. **[4 marks]**

4 Given that $\left(\sqrt[p]{x^{2p}}\right)^q = x^p \times x^{q+1}$, find an expression for p in terms of q **[4 marks]**

5   A point P lies on the curve with equation $y = x^2 - 2$

The normal to the curve at P is parallel to the line with equation $x = -2y$

Find the equation of the tangent to the curve at P **[6 marks]**

6 a   Describe the single geometrical transformation which maps $y=\sqrt{x}$ onto $y=3+\sqrt{x-4}$ **[2 marks]**

---

b i   State the domain and range of the function $f(x)=3+\sqrt{x-4}$ **[2 marks]**

---

ii   Hence sketch the graph of $y=f(x)$ on the set of axes below.   **[2 marks]**

6 c i f(x) has an inverse function f$^{-1}(x)$

Find f$^{-1}(x)$ and state its domain and range.   **[6 marks]**

6　c　ii　Sketch the graph of $y = f^{-1}(x)$ on the set of axes below.　**[2 marks]**

7 a   Find the range of values of x for which the curve $f(x) = x^3 - 12x + 3$ is decreasing.   **[3 marks]**

b   Use the second derivative test to confirm that $f(x)$ has an inflection point at $(0, 3)$ **[3 marks]**

8   Given that θ is measured in radians, use the method of **differentiation by first principles** to find the derivative of the function $f(\theta) = \cos\theta$ with respect to θ

You should show all your working, but you may assume the small angle approximations for $\sin A$ and $\cos A$, and the compound angle formula for $\cos(A + B)$ **[6 marks]**

9 If $\dfrac{dy}{dx} = \dfrac{1}{x^2 - x}$, find an expression for y in terms of x

Give your answer as a **single** logarithmic term.   **[8 marks]**

10   A temperature of 50° Fahrenheit is equal to 10° Celsius.

A temperature of 86° Fahrenheit is equal to 30° Celsius.

a   Find an equation relating Fahrenheit, F, to Celsius, C.

Give your answer in the form $C = pF + q$, where p and q are rational numbers.   **[4 marks]**

b   Interpret, in context, the values of p and q **[2 marks]**

10 c What temperature, in degrees Celsius, is exactly half the equivalent temperature in degrees Fahrenheit? **[2 marks]**

11 a Find the *exact* solution to the inequality $\frac{1}{2}\left(\frac{3}{4}-\frac{x}{2}\right)-\frac{3}{8}\left(\frac{3x}{5}-\frac{1}{3}\right)<0$

You must show all your working. **[3 marks]**

11 b Find the *exact* solution to the inequality $2x^2 - \dfrac{27x}{8} - \dfrac{11}{3} \ge \dfrac{3x}{27} + \dfrac{1}{12}$

You must show all your working. **[5 marks]**

c Solve the inequality $1 < x^2 < 4$ **[3 marks]**

12 a Complete the rows in the table below.

Give each value correct to six decimal places.

x	2	3	5	6	8
y	7.68	6.144	3.93216	3.145728	2.013 265 92
$\log_{10} x$					
$\log_{10} y$					

[1 mark]

b i Sketch the graph of $\log_{10} y$ against $\log_{10} x$ [2 marks]

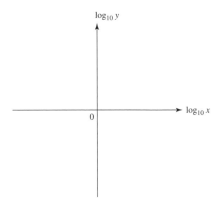

ii Sketch the graph of $\log_{10} y$ against x [2 marks]

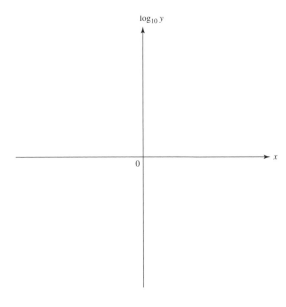

12 c i Determine whether the relationship between the data in the table is of the form $y = ax^b$, or of the form $y = ab^x$

Justify your answer. **[3 marks]**

_____

_____

_____

_____

_____

_____

_____

_____

_____

_____

ii Find the values of the constants a and b

Give the value of a correct to 2 significant figures, and the value of b correct to 1 significant figure. **[6 marks]**

_____

_____

_____

_____

_____

_____

_____

_____

_____

_____

_____

_____

_____

_____

_____

_____

_____

13 Prove that $\dfrac{\sin^2 2x}{4} + \sin^4 x = 1 - \dfrac{\cot^2 x}{\operatorname{cosec}^2 x}$ **[6 marks]**

14   The curve defined by the parametric equations $x = t(t^2 - 4)$, $y = 6(t^2 - 4)$ is sketched below.

a   At what point does the curve cross itself?

Justify your answer.   **[3 marks]**

14 b Find the equations of both tangents to the curve at the point where the curve crosses itself. **[6 marks]**

End of questions

| Name | | Class | |

| Signature | | Date | |

Candidates may use any calculator permitted by ICE regulations. Calculators must not have the facility for algebraic manipulation, differentiation and integration, or have retrievable mathematical formulae stored in them.

Question	Mark
1	
2	
3	
4	
5	
6	
7	
8	
9	
10	
11	
12	
Total	

Instructions

- Use black ink or ball-point pen.
- If pencil is used for diagrams/sketches/graphs it must be dark (HB or B).
- **Fill in the boxes** at the top of this page with your name and class.
- Answer all the questions and ensure that your answers to parts of questions are clearly labelled.
- Answer the questions in the spaces provided
  – *there may be more space than you need.*
- You should show sufficient working to make your methods clear. Answers without working may not gain full credit.
- Inexact answers should be given to three significant figures unless otherwise stated.

Information

- A booklet 'Mathematical Formulae and Statistical Tables' is provided.
- There are 12 questions in this question paper. The total mark for this paper is 100.
- The marks for each question are shown in brackets
  – *use this as a guide as to how much time to spend on each question.*

Advice

- Read each question carefully before you start to answer it.
- Try to answer every question.
- Check your answers if you have time at the end.

Section A

Answer **all** questions in the spaces provided.

1 Events A and B are not independent and are not mutually exclusive.

If $P(A) = 0.4$ and $P(B) = 0.7$, what is $P(A')$?   **[1 mark]**

_____

_____

_____

_____

_____

2 The standard Normal density function describes a probability distribution and has a bell-shaped curve. It is usually written as $Z \sim N(0, 1)$

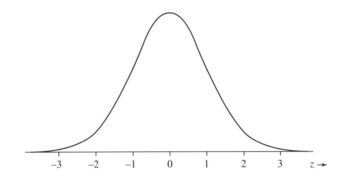

Use the fact that $P(Z < -0.4) = 0.3446$ and $P(Z < 0.2) = 0.5793$ to find the following to 3 significant figures without using your calculator.

You must show all your working.

a $P(Z > 0.2)$ **[2 marks]**

_____

_____

_____

b $P(Z < 0.4)$ **[2 marks]**

_____

_____

_____

c $P(-0.4 < Z < 0.2)$ **[2 marks]**

_____

_____

_____

3   A bag contains a large number of balls. Each ball is either red, blue or green, and is either spotted or plain. The numbers of each type are given in this table.

	Red	Blue	Green
Spotted	34	41	75
Plain	27	23	50

a   A single ball is drawn from the bag at random. Show your working and calculate the probability that the drawn ball

   i   Is green and spotted,   **[1 mark]**

   ii   Is blue or plain,   **[1 mark]**

   iii   Is red, given that it is spotted.   **[2 marks]**

b   Are the events 'the ball is green' and 'the ball is plain' independent? Explain your answer.   **[2 marks]**

4 The probability distribution for a random variable X is given by

$$P(X=x) = \frac{k}{1+2(x-2)^2} \text{ for } x = 0, 1, 2, 3, 4$$

where k is a rational number.

a In terms of k, Find the probability of each value that X can take, and hence calculate the value of k **[4 marks]**

X is used to approximate a Normally distributed variable Y with mean 2 and standard deviation 0.75

b i Explain why $P(1.5 < Y < 2.5) \approx P(X=2)$ **[1 mark]**

    ii Calculate the probabilities of obtaining outcomes which round to each of 0, 1, 2, 3, and 4 with this Normal distribution.   **[3 marks]**

4 c Using your answers to **a** and **b**, comment on the validity of using X to
approximate Y **[2 marks]**

_____

_____

_____

_____

5   The following is an excerpt from the Met Office's website regarding their sampling for the Large Data Set.

'The Met Office has a weather station network across the whole UK, with more than 200 automatic stations.

These weather stations measure a large variety of different meteorological parameters, including air temperature; atmospheric pressure; rainfall; wind speed and direction, humidity; cloud height and visibility.

Stations are usually around 40 km apart, enabling us to record the weather associated with the typical low pressure and frontal systems that cross the UK.'

a   Explain the meaning of the terms systematic sample and stratified sample.   **[4 marks]**

b   Define the two sampling frames used.   **[1 mark]**

c   Explain how to take a systematic sample of data from weather stations on a single day.   **[2 marks]**

Lucas is investigating how the temperature has changed in Leuchars over time.

He takes data from the Large Data Set for 1987 and 2015 and calculates the differences between the average temperature on consecutive days in July.

Lucas presents his findings using the table and histogram shown below.

The raw data acquired by Lucas (showing differences between the average temperature on consecutive days in July) is shown in ascending order below.

−5.7°C	−5.3°C	−5.1°C	−4.7°C	−4.6°C	−4.1°C	−3.7°C
−2.6°C	−2.4°C	−2.2°C	−2.2°C	−2.2°C	−2.1°C	−1.7°C
−1.7°C	−1.4°C	−1°C	−0.6°C	−0.6°C	−0.5°C	−0.5°C
0.3°C	0.8°C	0.9°C	1.8°C	2.1°C	2.2°C	3.5°C
3.6°C	4.9°C	5.2°C				

Difference between average temperatures on consecutive days (°C)	−6 °C to −5 °C	−5 °C to −2.5 °C	−2.5 °C to −1 °C	−1 °C to 0 °C	0 °C to 2 °C	2 °C to 4 °C	4 °C to 6 °C
Frequency			8	5	4	4	2

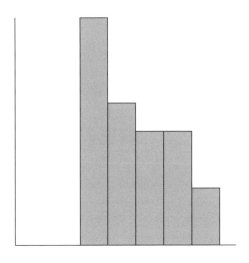

a

   i   Label the axes of the histogram.   **[1 mark]**

   ii   Use the histogram to fill in the missing values in the table.   **[2 marks]**

   iii   Add the two missing bars to the histogram using the table of values.   **[2 marks]**

Christine says that the data collected by Lucas looks like it could be drawn from a Normal distribution.

6 b   Give two features of Lucas's data which support Christine's claim, and one feature of the data which might suggest that Christine's claim is wrong.   **[3 marks]**

c   Calculate the mean and variance of the sample.   **[2 marks]**

The conversion of temperature from Celsius C to Fahrenheit F is $F = 1.8C + 32$

d   Use your answer to **c** to calculate the mean and variance of the temperature differences in Fahrenheit.   **[3 marks]**

Lucas collected the data in order to test the hypothesis that there was no difference between the average temperature on consecutive days in July.

He assumes that the differences can be modelled by a Normal distribution with variance equal to the sample variance in **c**

6   e   Perform Lucas's hypothesis test at the 5% level, stating the hypotheses and conclusion clearly.   **[7 marks]**

End of Section A

Section B

Answer **all** questions in the spaces provided.

7   Find the resultant of the vectors $4\mathbf{i} - 3\mathbf{j}$ and $7\mathbf{i} + \mathbf{j}$ **[1 mark]**

_____

_____

8   A uniform rod AB of length 3 m and mass 6 kg is pivoted smoothly at A

The rod is held in equilibrium at an angle of 45° to the horizontal by a force $\mathbf{F}$ which acts perpendicular to the rod at B, as shown in the diagram.

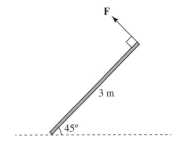

Calculate the magnitude of $\mathbf{F}$ **[4 marks]**

_____

_____

_____

_____

_____

_____

_____

_____

_____

_____

_____

_____

9   A block of mass 12 kg rests on a surface which is inclined at 40° to the horizontal.

A light string is attached to the top corner of the block and is held taut at an angle of 30° above the slope. The string holds the block in limiting equilibrium so that the block is about to slide down the slope. The coefficient of friction between the block and the slope is 0.5

By first drawing a diagram to show all the forces acting on the block, find the normal reaction **R** and the tension in the string **T**   **[7 marks]**

10   A ball is thrown from a window which is 12 m above horizontal ground.

The initial speed of the ball is 15 ms⁻¹ and it is thrown at an angle of 10 ° below the horizontal.

The ball strikes a vertical wall that lies a horizontal distance of 9 m away from the window, as shown in the diagram.

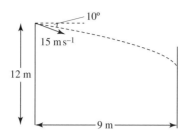

a   Calculate the height at which the ball strikes the wall.   **[5 marks]**

10 b Calculate the speed of the ball and its direction of motion immediately before it hits the wall. **[7 marks]**

11 A skydiver of mass 71 kg steps out of a hot-air balloon which is stationary and high above the ground.

Her velocity is $v\,\mathrm{ms^{-1}}$ at time t s after she leaves the balloon.

While falling, she experiences a resistive force (due to air resistance) of $5.8v$ N

a Draw a diagram to show the forces acting on the skydiver.   **[2 marks]**

b The skydiver accelerates towards the ground until the resistive force is equal to her weight.

The velocity at which she travels when these two forces are equal is called her *terminal velocity*.

Calculate the terminal velocity of the skydiver.   **[2 marks]**

c Use Newton's second law to show that

$$120 - v = 12.2\frac{\mathrm{d}v}{\mathrm{d}t}$$

Give your working to 3 sf.   **[3 marks]**

11 d   Find t when $v = 90 \text{ m s}^{-1}$ **[8 marks]**

12   Particles P and Q, with masses 0.5 kg and 0.3 kg respectively, are attached to the ends of a light, inextensible string that passes over a smooth light pulley.

The string is taut, and P rests in limiting equilibrium on a rough plane which is inclined at 55° to the horizontal, as shown in the diagram. P is on the point of slipping down the plane.

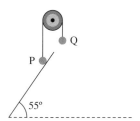

a   Calculate the coefficient of friction, μ, between P and the plane.   **[6 marks]**

12 b   Another particle of mass 0.3 kg is attached to Q.

The system is then released from rest.

Calculate the tension in the string and the acceleration of the particles.   **[5 marks]**

End of questions

A Level Mathematics
Paper 1 (Set C)

Edexcel

Name		Class	
Signature		Date	

Materials

For this paper you must have:

- The booklet of formulae and statistical tables
- You may use a graphics calculator.

Instructions

- Use black ink or black ball-point pen.
  Pencil should be used for drawing.
- Answer **all** questions.
- You must answer each question in the space provided for that question. If you require extra space, use a supplementary answer book; do not use the space provided for a different question.
- Do not write outside the box around each page.
- Show all necessary working; otherwise marks for method may be lost.
- Do all rough work in this book. Cross through any work that you do not want to be marked.

Question	Mark
1	
2	
3	
4	
5	
6	
7	
8	
9	
10	
11	
Total	

Information

- The marks for questions are shown in brackets.
- The maximum mark for this paper is 100.

Advice

- Unless stated otherwise, you may quote formulae, without proof, from the booklet.
- You do not necessarily have to use all the space provided.

Answer **all** questions in the spaces provided.

1 Write $\ln(2e^x)$ in the form $a + \ln b$, where a and b are integers. **[2 marks]**

2 Simplify $\dfrac{1}{\dfrac{2}{\dfrac{3}{x}+1}+1}$ **[3 marks]**

3 **a** Mark believes there is a solution to the following integration problem:

$$\int_{-2}^{3} \frac{1}{x^2} \, dx$$

Sally disagrees and says that this problem has no solution.

Who is correct? Give two reasons to justify your answer. **[2 marks]**

b To try to prove that he is correct, Mark offers the following solution to Sally:

$$\int_{-2}^{3} \frac{1}{x^2} \, dx = \left[-\frac{1}{x} \right]_{-2}^{3} = -\frac{1}{3} - \frac{1}{2} = -\frac{5}{6}$$

By considering the sign of $\frac{1}{x^2}$ and the sign of Mark's final answer, describe how Sally might know immediately that Mark's solution must be incorrect. **[3 marks]**

4 **a** **i** Use the trapezium rule, with 8 strips, to estimate $\int_{2}^{4} f(x)dx$ where

$$f(x) = \frac{(x+3)^2}{x-1}, \ x \neq 1$$

Give your estimate to 3 decimal places. **[5 marks]**

_____

_____

_____

_____

_____

_____

_____

_____

_____

_____

_____

_____

_____

_____

_____

_____

_____

ii With the aid of an appropriate sketch, state whether your estimate in **4 a i** is likely to be an overestimate or an underestimate of the actual value. Give a reason for your answer.

[3 marks]

4 b i Show that $(x+3)^2$ can be written as $x(x-1)+a(x-1)+b$, where a and b are integers to be found.

[3 marks]

ii Use your answer to **4 b i**, along with a suitable method of integration, to find the *exact* value of

$$\int_2^4 f(x)\,dx$$

You must show every step of your working to gain full marks. **[5 marks]**

iii Hence calculate, to the nearest 0.1%, the percentage error in your estimate of $\int_2^4 f(x)\,dx$ which you found in **4 a i** by using the trapezium rule. **[2 marks]**

5 Given that θ is small and measured in radians, find *exact* approximations for the following expressions

a $\dfrac{\cos 4\theta - 1}{\sin^2 3\theta}$

[3 marks]

b $\dfrac{4\cos 2\theta + \theta \sin^3 \theta}{\sin^4 \theta - 2\sin^2 2\theta + 4}$

[3 marks]

5 c $\dfrac{1-\sin\left(\dfrac{\pi}{2}-\theta\right)}{\cos^2\left(\dfrac{\pi}{2}-\theta\right)}$ **[3 marks]**

6 **a** Show that the circles $x^2 + y^2 + 8x - 6y - 11 = 0$ and $x^2 + y^2 - 16x - 16y + 79 = 0$ touch each other.

[7 marks]

6 b i Show that the two circles touch at the point with coordinates $\left(\dfrac{20}{13}, \dfrac{69}{13}\right)$ **[3 marks]**

ii Find the equation of the line which is tangent to both circles at $\left(\dfrac{20}{13}, \dfrac{69}{13}\right)$

You must state any reasoning or assumptions that you use. **[4 marks]**

7 The rate of growth of a population P at time t is directly proportional to the total population at time t days

a Write down a differential equation relating P and t [1 mark]

b By solving the differential equation, show that the general solution of the equation can be written in the form $P = Ae^{kt}$, where A and k are constants. [3 marks]

7 c Initially, the population is 5.2 million. 14 days later, it has grown to 6 million.

Find the exact values of A and k

Find also the size of the population after a *further* 30 days.

Give your answer correct to 3 significant figures. **[5 marks]**

8   The ellipse below is defined by the parametric equations $x = 5\cos t - 2$, $y = 2\sin t - 1$, where t is a parameter.

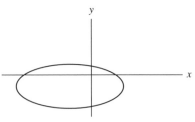

Find a Cartesian equation for the ellipse in the form $a(x+b)^2 + c(y+d)^2 = e$, where a, b, c, d and e are all positive integers.

[3 marks]

9 a Use the substitution $u = \cos x$ to show that $\int \tan x \ \mathrm{d}x = \ln|\sec x| + c$, where c is an arbitrary constant.

Your working should be both clear and rigorous. **[6 marks]**

9 b Use a suitable trigonometric substitution to show that $\int \sec^4 x \tan x \; \mathrm{d}x = \dfrac{\sec^4 x}{4} + C$, where C is an arbitrary constant.

[5 marks]

9 c It is given that $\int \sec x \, dx = \ln|\sec x + \tan x| + K$, where K is an arbitrary constant.

Use this result, as well as your answers to **a** and **b**, to show that

$$\int \left[\sec x \tan x (\sec^3 x) - \sec x + \tan x \right] dx = \frac{\sec^4 x - 1}{4} + \ln\left(\frac{A \sec x}{\sec x + \tan x} \right)$$

where A is an arbitrary constant.   **[4 marks]**

10 a   Show that $10x^2 - 11x + 1$ is a quadratic factor of the quartic expression

$20x^4 - 72x^3 + 87x^2 - 38x + 3$ **[3 marks]**

b   Hence write $20x^4 - 72x^3 + 87x^2 - 38x + 3$ as a product of four linear factors.   **[2 marks]**

10 c You are given that $y = 20x^4 - 72x^3 + 87x^2 - 38x + 3$ has a turning point when $x \approx 0.35$ and another turning point when $x \approx 1.35$

Sketch the graph of y below, clearly indicating any points where your graph meets the coordinate axes.

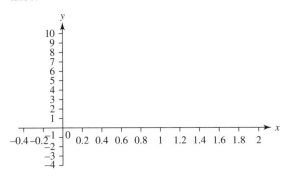

[2 marks]

_____

_____

_____

_____

_____

_____

_____

_____

_____

_____

_____

_____

d Describe the geometrical transformation that will map y onto

$20x^4 + 72x^3 + 87x^2 + 38x + 3$

[1 mark]

_____

_____

_____

_____

_____

_____

_____

_____

_____

_____

_____

11 a i Show that $\dfrac{3}{n^2}\left[\dfrac{n(n+1)}{2}\right]+4=\dfrac{11}{2}+\dfrac{3}{2n}$

[2 marks]

ii Use the formula $S_n=\dfrac{n}{2}\left[2a+(n-1)d\right]$ to show that the arithmetic series given by

$$\sum_{i=1}^{n}i=1+2+3+...+n \text{ has a sum of } S_n=\dfrac{n(n+1)}{2}$$

[2 marks]

11 b The definite integral $\int_a^b f(x)\,dx$, where $f(x)$ is a linear function, can be defined by

$$\int_a^b f(x)\,dx = \lim_{n\to\infty}\sum_{i=1}^{n}\left[\left(\frac{b-a}{n}\right)f(c_i)\right]$$

where $c_i = a + \left(\frac{b-a}{n}\right)i$

i Verify that, for the definite integral $\int_0^1 (3x+4)\,dx$, $c_i = \dfrac{i}{n}$ [1 mark]

ii Use the Fundamental Theorem of Calculus to evaluate $\int_0^1 (3x+4)\,dx$ [3 marks]

11 b iii Use the definition given at the beginning of the question to evaluate $\int\limits_0^1 (3x+4)\,\mathrm{d}x$

In your working, you may assume that $\sum\limits_{i=1}^n \dfrac{i}{n} = \dfrac{1}{n}\sum\limits_{i=1}^n i$ etc.   **[6 marks]**

End of questions

A Level Mathematics
Paper 2 (Set C)

Edexcel

| Name | | Class | |
| Signature | | Date | |

Materials

For this paper you must have:

- The booklet of formulae and statistical tables
- You may use a graphics calculator.

Instructions

- Use black ink or black ball-point pen.
  Pencil should be used for drawing.
- Answer **all** questions.
- You must answer each question in the space provided for that question. If you require extra space, use a supplementary answer book; do **not** use the space provided for a different question.
- Do not write outside the box around each page.
- Show all necessary working; otherwise marks for method may be lost.
- Do all rough work in this book. Cross through any work that you do not want to be marked.

Information

- The marks for questions are shown in brackets.
- The maximum mark for this paper is 100.

Advice

- Unless stated otherwise, you may quote formulae, without proof, from the booklet.
- You do not necessarily have to use all the space provided.

Question	Mark
1	
2	
3	
4	
5	
6	
7	
8	
9	
10	
11	
12	
13	
14	
15	
16	
Total	

Answer **all** questions in the spaces provided.

1   What is the range of validity of the binomial expansion of $\dfrac{1}{ax+b}$, where a and b are positive constants? **[1 mark]**

2   Find the general solution of the differential equation $\dfrac{dx}{dt} = kx$, where k is a constant. **[2 marks]**

3   **a**   Use the **product rule** to differentiate $\dfrac{e^{\frac{1}{2}x}}{2x}$ with respect to x [3 marks]

b   Use a suitable **substitution** to integrate $\dfrac{1}{x\ln x}$ with respect to x [4 marks]

4   The trapezium *ABCD* (shown below) has an area of $9(4\sqrt{3}+1)$ square units and a height of $2\sqrt{3}$ units.

The parallel sides *AD* and *BC* are such that $AD : BC = 2 : 1$

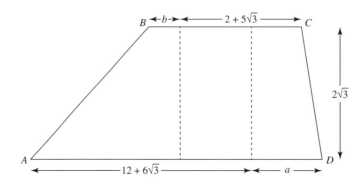

Find the exact lengths *a* and *b*   **[6 marks]**

---

5 Prove that the value of the expression

$$2x^2 - 12x + 19$$

is positive for all real values of x **[4 marks]**

6 a The cubic equation $x^3 - x^2 + 3x - 4 = 0$ has a single root, α

Show that α lies between $x = 1$ and $x = 1.5$ **[3 marks]**

6 b Show that the cubic equation can be rearranged into the form $x = \sqrt{\dfrac{x^2 + 4}{x} - 3}$ **[2 marks]**

c Using the rearrangement given in part **b**, state a recurrence relation that you could use to find α

Using $x_1 = 1.25$, use your relation to find α correct to two decimal places.

You must show all your working. **[5 marks]**

6 d The axes below show the curve $y = \sqrt{\dfrac{x^2+4}{x}} - 3$ and the line $y = x$ for $0 \le x \le 2$

On the graph, draw a cobweb or staircase diagram, with $x_1 = 1.25$, to show how convergence to α takes place via the recurrence relation in **c**.

Indicate on your diagram the positions of x_2 and x_3

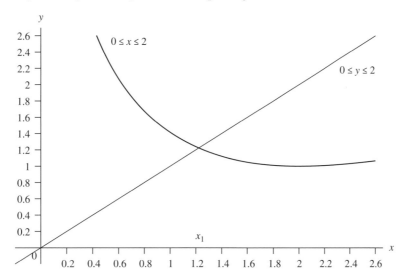

[4 marks]

7 By implicitly differentiating both sides of $(x+\sin^2 y)^2 = (xy)^2$, find an expression for $\dfrac{dy}{dx}$ **[5 marks]**

8 a i   What is the period, in radians, of the graph of $y = \cos bx$, where b is positive constant?   **[1 mark]**

_____

_____

ii   What is the amplitude of the graph of $y = \cos bx$?   **[1 mark]**

_____

_____

b i   What is the period, in radians, of the graph of $y = a\cos bx$, where a and b are both positive constants?   **[1 mark]**

_____

_____

ii   What is the amplitude of the graph of $y = a\cos bx$?   **[1 mark]**

_____

_____

c i   What is the period, in radians, of the graph of $y = a\cos bx + c$, where a, b and c are all positive constants?   **[1 mark]**

_____

_____

ii   What is the amplitude of the graph of $y = a\cos bx + c$?   **[1 mark]**

_____

_____

8   d   The graph below shows the number of hours daylight experienced by a city over the course of a year.

The longest day occurs on June 21st, when the city experiences 17 hours of daylight.

The shortest day is December 21st, when there are only 7.5 hours of daylight.

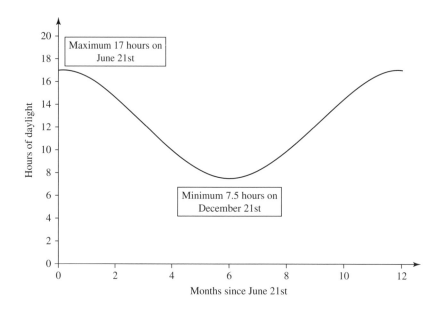

Write a trigonometric model (in radians) for the number of hours of daylight (y) in terms of the number of months since June 21st (x)

Give your model in the form $y = a\cos bx + c$, where a, b and c are positive constants.   **[7 marks]**

9   If $x = \cos^2 y$, find $\dfrac{\mathrm{d}y}{\mathrm{d}x}$ in terms of a trigonometric function of $2y$ **[5 marks]**

10   Blaise solves the inequality $x^2 < a^2$ for x, where a is a real number.

He writes his solution as

$$\{x : -a < x\} \cap \{x : x < a\}$$

Is Blaise correct? Explain your answer.   **[2 marks]**

11 **a** On the same set of axes below, sketch the graphs of $y = |4x - 1|$ and $y = |3 - x|$

Label clearly any points where the graphs meet the coordinate axes.

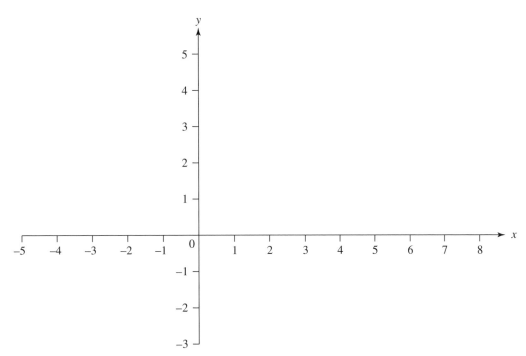

[4 marks]

_____

_____

11 b Find the exact values of the two pairs of coordinates where the graphs of $y=|4x-1|$ and $y=|3-x|$ intersect. **[6 marks]**

c Hence, write down the exact solution to the inequality $|3-x|>|4x-1|$ **[1 mark]**

12   The denominator of a fraction is five more than its numerator.

Both numerator and denominator are positive.

When four is added to both the numerator and denominator, the value of the fraction increases by $\dfrac{5}{24}$

What is the original fraction?

You must show all your working.   **[6 marks]**

13 Vince chooses any two integers between 0 and 9 (e.g. 4 and 5).

He forms a two-digit number using these two integers (e.g. 45). He then reverses the digits to form a second two-digit number (e.g. 54).

Vince observes that the sum of both two-digit numbers seems always to be a multiple of 11 (e.g. $45 + 54 = 99 = 9 \times 11$).

Prove that Vince's observation will always be true for **any** two-digit number.   **[4 marks]**

14 Describe the single geometrical transformation which maps

$f(x) = \cos 2x$ onto $g(x) = \cos\left(2x + \dfrac{\pi}{6}\right)$ **[2 marks]**

15 a Write down the three terms (*not* the sum) of the geometric sequence

represented by $\displaystyle\sum_{r=1}^{3} 2\left(\frac{\sqrt{2}}{2}\right)^{r-1}$ **[1 mark]**

b Find $\displaystyle\sum_{r=10}^{\infty} 2\left(\frac{\sqrt{2}}{2}\right)^{r-1}$

Write your answer in the form $a + b\sqrt{2}$, where a and b are rational numbers.

You must show all your working. **[6 marks]**

16 a   It is given that $\sin(A-B) \equiv \sin A \cos B - \cos A \sin B$

By replacing A with $\dfrac{\pi}{2} - A$, prove that $\cos(A+B) \equiv \cos A \cos B - \sin A \sin B$ **[3 marks]**

b   Use the result in part **a** to deduce that $\cos^2 A \equiv \dfrac{1+\cos 2A}{2}$ **[3 marks]**

16 c   **Hence,** find $\int \cos^4 A \, \mathrm{d}A$ **[5 marks]**

End of questions

A Level Mathematics
Paper 3 (Set C)

Edexcel

| Name | _____ | Class | _____ |
| Signature | _____ | Date | _____ |

Candidates may use any calculator permitted by ICE regulations. Calculators must not have the facility for algebraic manipulation, differentiation and integration, or have retrievable mathematical formulae stored in them.

Instructions

- Use black ink or ball-point pen.
- If pencil is used for diagrams/sketches/graphs it must be dark (HB or B).
- **Fill in the boxes** at the top of this page with your name and class.
- Answer all the questions and ensure that your answers to parts of questions are clearly labelled.
- Answer the questions in the spaces provided
  – *there may be more space than you need.*
- You should show sufficient working to make your methods clear. Answers without working may not gain full credit.
- Inexact answers should be given to three significant figures unless otherwise stated.

Question	Mark
1	
2	
3	
4	
5	
6	
7	
8	
9	
10	
11	
12	
13	
Total	

Information

- A booklet 'Mathematical Formulae and Statistical Tables' is provided.
- There are 13 questions in this question paper. The total mark for this paper is 100.
- The marks for each question are shown in brackets
  – *use this as a guide as to how much time to spend on each question.*

Advice

- Read each question carefully before you start to answer it.
- Try to answer every question.
- Check your answers if you have time at the end.

Section A

Answer **all** questions in the spaces provided.

1   Events *A* and *B* are independent events. Explain what is meant by independent.   **[1 mark]**

_____

_____

_____

_____

2   A group of 16 students each take both a spelling test and a mental arithmetic test. Each test is marked out of a total of 10 marks. For the 16 students, the scores of both tests are presented in this scatter graph.

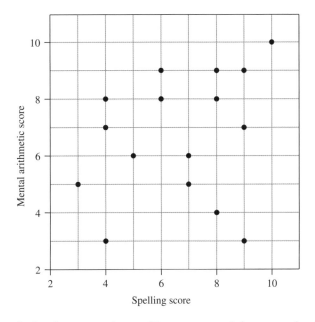

a   Describe the correlation between the spelling score and the mental arithmetic score.   **[1 mark]**

_____

_____

_____

2 b   The correlation coefficient for the data is 0.268 4

Use this data to perform a hypothesis test at the 5% significance level to decide if there is any correlation between students' scores in the spelling test and the mental arithmetic test.

The critical values are ±0.514   **[5 marks]**

3   When an unbiased, six-sided dice is rolled, the outcome is a face showing a
number between 1 and 6.

The events $E = \{2, 4, 6\}$ and $O = \{1, 3, 5\}$ are exhaustive.

a   State what is meant by the term exhaustive.   **[1 mark]**

The dice is rolled 4 times. The event $EEEO$ shows that the first three results are in the
set E and the fourth result is in the set O.

b   List all 16 possible outcomes from the 4 rolls.   **[2 marks]**

c   State the probability that a random outcome contains at least three results
from set E **[1 mark]**

d   Find the probability that a random outcome contains at least three results
from set E, given that it contains at least two results from set E **[1 mark]**

4   A polling company wants to see if a newly announced policy by a parliamentary party is popular within a constituency. They assume that one third of all people will agree with the policy and the rest will disagree with it.

The company takes a sample of size 30 by inviting people from the constituency to attend a lunchtime meeting in exchange for a small monetary gift.

a   State one potential weakness of this sampling method.   **[1 mark]**

Let the random variable X represent the number of people in the sample who agree with the policy.

b   State the distribution of X **[1 mark]**

c   State the mean and variance of X **[2 marks]**

d   Using your distribution in **b**, find the probability that at most 5 people in the sample agree with the policy.   **[1 mark]**

After the meeting, it turned out that half the group agreed with the policy.

4 e Assuming that the group forms a random sample, perform a hypothesis test at the 5% level to decide if the company's assumption about the level of the support for the policy was an underestimate. **[7 marks]**

5 Kiran is investigating wind speeds. He collects data on the daily maximum gust (in knots) over a six-month period. He groups the data, and presents his findings using a cumulative frequency table and graph.

Speed range (knots)	Cumulative Frequency
0 – 8	0
9 – 12	9
13 – 15	25
16 – 18	51
19 – 20	68
20 – 21	81
22 – 24	114
25 – 27	148
28 – 30	169
31 – 36	184
37 – 40	191
41 – 43	194

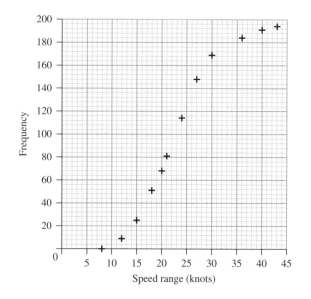

a Estimate the median daily maximum gust. **[1 mark]**

b Estimate the interquartile range (IQR) of daily maximum gusts. **[3 marks]**

5 c   Estimate the probability that for a randomly chosen day, the maximum gust
exceeds 30 knots.   **[2 marks]**

_____

_____

_____

_____

Kieran also collects data on the daily maximum gust for a different town over the same
six-month period.

His results for this town are shown in the cumulative frequency table below.

Speed range (knots)	Cumulative Frequency
0 – 7	0
8 – 12	9
13 – 15	29
16 – 18	65
19 – 20	90
20 – 21	102
22 – 24	142
24 – 27	171
28 – 30	180
31 – 36	189
37 – 40	191
41 – 43	192

d   Draw a second cumulative frequency curve on the graph above to represent
this data.   **[3 marks]**

e   Compare the two distributions, referring to the central tendency and the
spread.   **[2 marks]**

_____

_____

_____

_____

_____

_____

_____

_____

6 Eashan is investigating daily total sunshine levels across Britain.

He collects data over a period of six months for five different regions of England, and generates box-and-whisker plots from the summary statistics given in this table.

	Min	LQ	Med	UQ	Max
Camborne	0	1.1	4.9	8.8	15.4
Heathrow	0	1.2	4.8	8.1	15.1
Hurn	0	1.6	5.2	8.6	15
Leeming	0	0.5	2.8	6.5	13.6
Leuchars	0	1.3	4.6	8	13.3

a Draw the box-and-whisker plot for Hurn. **[3 marks]**

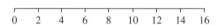

b State the region with the largest range and the region with the smallest range in daily total sunshine levels. **[2 marks]**

6 c   Does the data for Leeming contain any outliers? Justify your answer.   **[4 marks]**

Cluster sampling is to be used for further studies.

d   From which of the given weather stations would it be best to take a sample to get a reflection of the average across Britain? Justify your choice.   **[2 marks]**

e   State one advantage and one disadvantage of taking a sample in this way.   **[2 marks]**

f   Define the term simple random sampling and explain why cluster sampling does not provide a simple random sample.   **[2 marks]**

End of section A

Section B

Answer **all** questions in the spaces provided.

7   What feature of a velocity–time graph represents the displacement?   **[1 mark]**

_____

_____

8   Relative to perpendicular axes x and y, points A and B have
coordinates $(2, 6)$ and $(5, 2)$ respectively.

a   Find the acute angle which the line joining A and B makes with the x-axis.   **[2 marks]**

_____

_____

_____

_____

b   Find the unit vector in the direction of $\overrightarrow{AB}$

Give your answer in the form $a\mathbf{i} + b\mathbf{j}$, where $\mathbf{i}$ and $\mathbf{j}$ are the unit vectors in the

x- and y-directions respectively.   **[3 marks]**

_____

_____

_____

9   A uniform bar AB is 2.8 m long and weighs 80 N

Weights of 20 N and 40 N are attached to the bar at A and B respectively.

With the weights attached, the bar remains in horizontal equilibrium when supported at position C, which lies between A and B

a   Find the magnitude of the reaction at C **[2 marks]**

b   Find the distance AC **[3 marks]**

10 $\mathbf{i}$ and $\mathbf{j}$ are unit vectors in the x- and y- directions respectively.

Initially, a particle is at the point with position vector $\mathbf{r}_0 = -12\mathbf{i} + 2\mathbf{j}$ m and is travelling with velocity $\mathbf{v}_0 = (8\mathbf{i} - 3\mathbf{j})\ \mathrm{ms}^{-1}$

The particle accelerates uniformly for 8 seconds until it reaches the point with position vector $\mathbf{r} = 28\mathbf{i} + 10\mathbf{j}$ m

Find the acceleration of the particle.   **[4 marks]**

11 A force **F** acts on a particle of mass 6 kg, causing the particle to move in a straight line.

The particle's displacement, s, from the origin O at time t is found to be $s = te^{5-t}$ m

a Show that the velocity of the particle at time t will be given by $v = -e^{5-t}(t-1)$ ms^{-1} **[3 marks]**

b Find the magnitude of **F** when the particle when it is stationary. **[7 marks]**

11 b (cont.)

12   A box of mass 8 kg is held at rest on a slope which is inclined at 35° to the horizontal.

The box is connected to a mass of 2 kg by a light, inextensible string which passes over a smooth pulley. The 2 kg mass hangs freely below the pulley, as shown in the diagram.

The coefficient of friction between the box and the slope is 0.3, and the system is initially held at rest in equilibrium.

When the system is released from rest, it takes 3 seconds for the box to slide s m to the bottom of the slope.

Find the distance s **[8 marks]**

12 (cont.)

13 a   A particle is projected from ground level on a horizontal surface.

The initial speed of the particle is u ms^{-1} and it is projected at an angle of $\theta°$ to the horizontal.

After t seconds have elapsed, the particle is at the point (x, y)

Show that the particle moves in a path given by the equation $y = x\tan\theta - \dfrac{g\sec^2\theta}{2u^2}x^2$ **[7 marks]**

13 b   A waterskier performs a stunt where she takes off from a ramp at 9 ms^{-1} and tries to clear an obstacle that is 1m higher than the end of the ramp.

The obstacle lies in a vertical plane at a horizontal distance of 4 m from the end of the ramp.

Using the result from **a**, find the maximum and minimum angles that the ramp must make with the horizontal if the waterskier is to clear the obstacle.

You must show all your working.   **[6 marks]**

13 c Another model of the waterskier takes account of air resistance acting on the waterskier. How would you expect the maximum and minimum values of θ given by this model to be different from the values of θ you calculated in part **b**?

Give a reason for your answer. **[3 marks]**

End of questions

Mathematical formulae
For A Level Maths

The following mathematical formulae will be provided for you.

Pure Mathematics

Mensuration

Surface area of sphere $= 4\pi r^2$

Area of curved surface of cone $= \pi r \times$ slant height

Arithmetic series

$$S_n = \frac{1}{2}n(a+l) = \frac{1}{2}n[2a+(n-1)d]$$

Binomial series

$$(a+b)^n = a^n + \binom{n}{1}a^{n-1}b + \binom{n}{2}a^{n-2}b^2 + \cdots + \binom{n}{r}a^{n-r}b^r + \cdots + b^n \qquad (n \in \mathbb{N})$$

where $\binom{n}{r} = {}^nC_r = \dfrac{n!}{r!(n-r)!}$

$$(1+x)^n = 1 + nx + \frac{n(n-1)}{1 \times 2}x^2 + \cdots + \frac{n(n-1)\ldots(n-r+1)}{1 \times 2 \times \ldots \times r}x^r + \cdots \qquad (|x|<1, \ n \in \mathbb{N})$$

Logarithms and exponentials

$$\log_a x = \frac{\log_b x}{\log_b a}$$

$$e^{x\ln a} = a^x$$

Geometric series

$$S_n = \frac{a(1-r^n)}{1-r}$$

$$S_\infty = \frac{a}{1-r} \ \text{ for } |r|<1$$

Trigonometric identities

$$\sin(A \pm B) = \sin A \cos B \pm \cos A \sin B$$

$$\cos(A \pm B) = \cos A \cos B \mp \sin A \sin B$$

$$\tan(A \pm B) = \frac{\tan A \pm \tan B}{1 \mp \tan A \tan B} \qquad \left(A \pm B \neq \left(k + \frac{1}{2} \right)\pi \right)$$

$$\sin A + \sin B = 2\sin\frac{A+B}{2}\cos\frac{A-B}{2}$$

$$\sin A - \sin B = 2\cos\frac{A+B}{2}\sin\frac{A-B}{2}$$

$$\cos A + \cos B = 2\cos\frac{A+B}{2}\cos\frac{A-B}{2}$$

$$\cos A - \cos B = 2\sin\frac{A+B}{2}\sin\frac{A-B}{2}$$

Differentiation

First principles

$$f'(x) = \lim_{h \to 0} \frac{f(x+h) - f(x)}{h}$$

$f(x)$	$f'(x)$
$\tan kx$	$k \sec^2 kx$
$\sec kx$	$k \sec kx \tan kx$
$\cot kx$	$-k \csc^2 x$
$\csc kx$	$-\csc kx \cot kx$
$\dfrac{f(x)}{g(x)}$	$\dfrac{f'(x)g(x) - f(x)g'(x)}{(g(x))^2}$

Integration (+ constant)

$f(x)$	$\int f(x)\,\mathrm{d}x$	
$\sec^2 kx$	$\dfrac{1}{k}\tan kx$	
$\tan kx$	$\dfrac{1}{k}\ln\lvert\sec kx\rvert$	
$\cot kx$	$\dfrac{1}{k}\ln\lvert\sin kx\rvert$	
$\csc kx$	$-\dfrac{1}{k}\ln\lvert\csc kx + \cot kx\rvert,$	$\dfrac{1}{k}\ln\left\lvert\tan\left(\dfrac{1}{2}kx\right)\right\rvert$
$\sec kx$	$-\dfrac{1}{k}\ln\lvert\sec kx + \tan kx\rvert,$	$\dfrac{1}{k}\ln\left\lvert\tan\left(\dfrac{1}{2}kx + \dfrac{1}{4}\pi\right)\right\rvert$

$$\int u \frac{\mathrm{d}v}{\mathrm{d}x}\,\mathrm{d}x = uv - \int v \frac{\mathrm{d}u}{\mathrm{d}x}\,\mathrm{d}x$$

Numerical Methods

The trapezium rule: $\displaystyle\int_a^b y\,\mathrm{d}x \approx \frac{1}{2}h\{(y_0 + y_n) + 2(y_1 + y_2 + \ldots + y_{n-1})\}$, where $h = \dfrac{b-a}{n}$

The Newton-Raphson iteration for solving $f(x) = 0$: $x_{n+1} = x_n - \dfrac{f(x_n)}{f'(x_n)}$

Mechanics

Kinematics

For motion in a straight line with constant acceleration:

$$v = u + at$$
$$s = ut + \frac{1}{2}at^2$$
$$s = vt - \frac{1}{2}at^2$$
$$v^2 = u^2 + 2as$$
$$s = \frac{1}{2}(u+v)t$$

Mathematical formulae for A Level Maths

Statistics

Probability

$$P(A') = 1 - P(A)$$
$$P(A \cup B) = P(A) + P(B) - P(A \cap B)$$
$$P(A \cap B) = P(A)P(A|B)$$
$$P(A|B) = \frac{P(B|A)P(A)}{P(B|A)P(A) + P(B|A')P(A')}$$

For independent events A and B,
$$P(B|A) = P(B),$$
$$P(A|B) = P(A),$$
$$P(A \cap B) = P(A)P(B)$$

Standard deviation

Standard deviation $= \sqrt{\text{Variance}}$

Interquartile range $= \text{IQR} = Q_3 - Q_1$

For a set of n values $x_1, x_2, ..., x_i, ... x_n$

$$S_{xx} = \sum(x_i - \bar{x})^2 = \sum x_i^2 - \frac{\left(\sum x_i\right)^2}{n}$$

Standard deviation $= \sqrt{\dfrac{S_{xx}}{n}}$ or $\sqrt{\dfrac{\sum x^2}{n} - \bar{x}^2}$

Discrete distributions

Distribution of X	$P(X=x)$	Mean	Variance
Binomial $\quad$ B(n, p)	$\binom{n}{x}p^x(1-p)^x$	np	$np(1-p)$

Sampling distributions

For random sample of n observations from N(m, σ^2)

$$\frac{\bar{X} - \mu}{\sigma/\sqrt{n}} \sim \text{N}(0, 1)$$

The following statistical tables will be provided for you.

Binomial cumulative distribution function

The tabulated value is $P(X \le x)$, where X has a binomial distribution with index n and parameter p

$p =$	0.05	0.10	0.15	0.20	0.25	0.30	0.35	0.40	0.45	0.50
$n = 5, x = 0$	0.7738	0.5905	0.4437	0.3277	0.2373	0.1681	0.1160	0.0778	0.0503	0.0313
1	0.9774	0.9185	0.8352	0.7373	0.6328	0.5282	0.4284	0.3370	0.2562	0.1875
2	0.9988	0.9914	0.9734	0.9421	0.8965	0.8369	0.7648	0.6826	0.5931	0.5000
3	1.0000	0.9995	0.9978	0.9933	0.9844	0.9692	0.9460	0.9130	0.8688	0.8125
4	1.0000	1.0000	0.9999	0.9997	0.9990	0.9976	0.9947	0.9898	0.9815	0.9688
$n = 6, x = 0$	0.7351	0.5314	0.3771	0.2621	0.1780	0.1176	0.0754	0.0467	0.0277	0.0156
1	0.9672	0.8857	0.7765	0.6554	0.5339	0.4202	0.3191	0.2333	0.1636	0.1094
2	0.9978	0.9842	0.9527	0.9011	0.8306	0.7443	0.6471	0.5443	0.4415	0.3438
3	0.9999	0.9987	0.9941	0.9830	0.9624	0.9295	0.8826	0.8208	0.7447	0.6563
4	1.0000	0.9999	0.9996	0.9984	0.9954	0.9891	0.9777	0.9590	0.9308	0.8906
5	1.0000	1.0000	1.0000	0.9999	0.9998	0.9993	0.9982	0.9959	0.9917	0.9844
$n = 7, x = 0$	0.6983	0.4783	0.3206	0.2097	0.1335	0.0824	0.0490	0.0280	0.0152	0.0078
1	0.9556	0.8503	0.7166	0.5767	0.4449	0.3294	0.2338	0.1586	0.1024	0.0625
2	0.9962	0.9743	0.9262	0.8520	0.7564	0.6471	0.5323	0.4199	0.3164	0.2266
3	0.9998	0.9973	0.9879	0.9667	0.9294	0.8740	0.8002	0.7102	0.6083	0.5000
4	1.0000	0.9998	0.9988	0.9953	0.9871	0.9712	0.9444	0.9037	0.8471	0.7734
5	1.0000	1.0000	0.9999	0.9996	0.9987	0.9962	0.9910	0.9812	0.9643	0.9375
6	1.0000	1.0000	1.0000	1.0000	0.9999	0.9998	0.9994	0.9984	0.9963	0.9922
$n = 8, x = 0$	0.6634	0.4305	0.2725	0.1678	0.1001	0.0576	0.0319	0.0168	0.0084	0.0039
1	0.9428	0.8131	0.6572	0.5033	0.3671	0.2553	0.1691	0.1064	0.0632	0.0352
2	0.9942	0.9619	0.8948	0.7969	0.6785	0.5518	0.4278	0.3154	0.2201	0.1445
3	0.9996	0.9950	0.9786	0.9437	0.8862	0.8059	0.7064	0.5941	0.4770	0.3633
4	1.0000	0.9996	0.9971	0.9896	0.9727	0.9420	0.8939	0.8263	0.7396	0.6367
5	1.0000	1.0000	0.9998	0.9988	0.9958	0.9887	0.9747	0.9502	0.9115	0.8555
6	1.0000	1.0000	1.0000	0.9999	0.9996	0.9987	0.9964	0.9915	0.9819	0.9648
7	1.0000	1.0000	1.0000	1.0000	1.0000	0.9999	0.9998	0.9993	0.9983	0.9961
$n = 9, x = 0$	0.6302	0.3874	0.2316	0.1342	0.0751	0.0404	0.0207	0.0101	0.0046	0.0020
1	0.9288	0.7748	0.5995	0.4362	0.3003	0.1960	0.1211	0.0705	0.0385	0.0195
2	0.9916	0.9470	0.8591	0.7382	0.6007	0.4628	0.3373	0.2318	0.1495	0.0898
3	0.9994	0.9917	0.9661	0.9144	0.8343	0.7297	0.6089	0.4826	0.3614	0.2539
4	1.0000	0.9991	0.9944	0.9804	0.9511	0.9012	0.8283	0.7334	0.6214	0.5000
5	1.0000	0.9999	0.9994	0.9969	0.9900	0.9747	0.9464	0.9006	0.8342	0.7461
6	1.0000	1.0000	1.0000	0.9997	0.9987	0.9957	0.9888	0.9750	0.9502	0.9102
7	1.0000	1.0000	1.0000	1.0000	0.9999	0.9996	0.9986	0.9962	0.9909	0.9805
8	1.0000	1.0000	1.0000	1.0000	1.0000	1.0000	0.9999	0.9997	0.9992	0.9980

$p =$	0.05	0.10	0.15	0.20	0.25	0.30	0.35	0.40	0.45	0.50
$n = 10, x = 0$	0.5987	0.3487	0.1969	0.1074	0.0563	0.0282	0.0135	0.0060	0.0025	0.0010
1	0.9139	0.7361	0.5443	0.3758	0.2440	0.1493	0.0860	0.0464	0.0233	0.0107
2	0.9885	0.9298	0.8202	0.6778	0.5256	0.3828	0.2616	0.1673	0.0996	0.0547
3	0.9990	0.9872	0.9500	0.8791	0.7759	0.6496	0.5138	0.3823	0.2660	0.1719
4	0.9999	0.9984	0.9901	0.9672	0.9219	0.8497	0.7515	0.6331	0.5044	0.3770
5	1.0000	0.9999	0.9986	0.9936	0.9803	0.9527	0.9051	0.8338	0.7384	0.6230
6	1.0000	1.0000	0.9999	0.9991	0.9965	0.9894	0.9740	0.9452	0.8980	0.8281
7	1.0000	1.0000	1.0000	0.9999	0.9996	0.9984	0.9952	0.9877	0.9726	0.9453
8	1.0000	1.0000	1.0000	1.0000	1.0000	0.9999	0.9995	0.9983	0.9955	0.9893
9	1.0000	1.0000	1.0000	1.0000	1.0000	1.0000	1.0000	0.9999	0.9997	0.9990
$n = 12, x = 0$	0.5404	0.2824	0.1422	0.0687	0.0317	0.0138	0.0057	0.0022	0.0008	0.0002
1	0.8816	0.6590	0.4435	0.2749	0.1584	0.0850	0.0424	0.0196	0.0083	0.0032
2	0.9804	0.8891	0.7358	0.5583	0.3907	0.2528	0.1513	0.0834	0.0421	0.0193
3	0.9978	0.9744	0.9078	0.7946	0.6488	0.4925	0.3467	0.2253	0.1345	0.0730
4	0.9998	0.9957	0.9761	0.9274	0.8424	0.7237	0.5833	0.4382	0.3044	0.1938
5	1.0000	0.9995	0.9954	0.9806	0.9456	0.8822	0.7873	0.6652	0.5269	0.3872
6	1.0000	0.9999	0.9993	0.9961	0.9857	0.9614	0.9154	0.8418	0.7393	0.6128
7	1.0000	1.0000	0.9999	0.9994	0.9972	0.9905	0.9745	0.9427	0.8883	0.8062
8	1.0000	1.0000	1.0000	0.9999	0.9996	0.9983	0.9944	0.9847	0.9644	0.9270
9	1.0000	1.0000	1.0000	1.0000	1.0000	0.9998	0.9992	0.9972	0.9921	0.9807
10	1.0000	1.0000	1.0000	1.0000	1.0000	1.0000	0.9999	0.9997	0.9989	0.9968
11	1.0000	1.0000	1.0000	1.0000	1.0000	1.0000	1.0000	1.0000	0.9999	0.9998
$n = 15, x = 0$	0.4633	0.2059	0.0874	0.0352	0.0134	0.0047	0.0016	0.0005	0.0001	0.0000
1	0.8290	0.5490	0.3186	0.1671	0.0802	0.0353	0.0142	0.0052	0.0017	0.0005
2	0.9638	0.8159	0.6042	0.3980	0.2361	0.1268	0.0617	0.0271	0.0107	0.0037
3	0.9945	0.9444	0.8227	0.6482	0.4613	0.2969	0.1727	0.0905	0.0424	0.0176
4	0.9994	0.9873	0.9383	0.8358	0.6865	0.5155	0.3519	0.2173	0.1204	0.0592
5	0.9999	0.9978	0.9832	0.9389	0.8516	0.7216	0.5643	0.4032	0.2608	0.1509
6	1.0000	0.9997	0.9964	0.9819	0.9434	0.8689	0.7548	0.6098	0.4522	0.3036
7	1.0000	1.0000	0.9994	0.9958	0.9827	0.9500	0.8868	0.7869	0.6535	0.5000
8	1.0000	1.0000	0.9999	0.9992	0.9958	0.9848	0.9578	0.9050	0.8182	0.6964
9	1.0000	1.0000	1.0000	0.9999	0.9992	0.9963	0.9876	0.9662	0.9231	0.8491
10	1.0000	1.0000	1.0000	1.0000	0.9999	0.9993	0.9972	0.9907	0.9745	0.9408
11	1.0000	1.0000	1.0000	1.0000	1.0000	0.9999	0.9995	0.9981	0.9937	0.9824
12	1.0000	1.0000	1.0000	1.0000	1.0000	1.0000	0.9999	0.9997	0.9989	0.9963
13	1.0000	1.0000	1.0000	1.0000	1.0000	1.0000	1.0000	1.0000	0.9999	0.9995
14	1.0000	1.0000	1.0000	1.0000	1.0000	1.0000	1.0000	1.0000	1.0000	1.0000

$p=$	0.05	0.10	0.15	0.20	0.25	0.30	0.35	0.40	0.45	0.50
$n = 20, x = 0$	0.3585	0.1216	0.0388	0.0115	0.0032	0.0008	0.0002	0.0000	0.0000	0.0000
1	0.7358	0.3917	0.1756	0.0692	0.0243	0.0076	0.0021	0.0005	0.0001	0.0000
2	0.9245	0.6769	0.4049	0.2061	0.0913	0.0355	0.0121	0.0036	0.0009	0.0002
3	0.9841	0.8670	0.6477	0.4114	0.2252	0.1071	0.0444	0.0160	0.0049	0.0013
4	0.9974	0.9568	0.8298	0.6296	0.4148	0.2375	0.1182	0.0510	0.0189	0.0059
5	0.9997	0.9887	0.9327	0.8042	0.6172	0.4164	0.2454	0.1256	0.0553	0.0207
6	1.0000	0.9976	0.9781	0.9133	0.7858	0.6080	0.4166	0.2500	0.1299	0.0577
7	1.0000	0.9996	0.9941	0.9679	0.8982	0.7723	0.6010	0.4159	0.2520	0.1316
8	1.0000	0.9999	0.9987	0.9900	0.9591	0.8867	0.7624	0.5956	0.4143	0.2517
9	1.0000	1.0000	0.9998	0.9974	0.9861	0.9520	0.8782	0.7553	0.5914	0.4119
10	1.0000	1.0000	1.0000	0.9994	0.9961	0.9829	0.9468	0.8725	0.7507	0.5881
11	1.0000	1.0000	1.0000	0.9999	0.9991	0.9949	0.9804	0.9435	0.8692	0.7483
12	1.0000	1.0000	1.0000	1.0000	0.9998	0.9987	0.9940	0.9790	0.9420	0.8684
13	1.0000	1.0000	1.0000	1.0000	1.0000	0.9997	0.9985	0.9935	0.9786	0.9423
14	1.0000	1.0000	1.0000	1.0000	1.0000	1.0000	0.9997	0.9984	0.9936	0.9793
15	1.0000	1.0000	1.0000	1.0000	1.0000	1.0000	1.0000	0.9997	0.9985	0.9941
16	1.0000	1.0000	1.0000	1.0000	1.0000	1.0000	1.0000	1.0000	0.9997	0.9987
17	1.0000	1.0000	1.0000	1.0000	1.0000	1.0000	1.0000	1.0000	1.0000	0.9998
18	1.0000	1.0000	1.0000	1.0000	1.0000	1.0000	1.0000	1.0000	1.0000	1.0000
$n = 25, x = 0$	0.2774	0.0718	0.0172	0.0038	0.0008	0.0001	0.0000	0.0000	0.0000	0.0000
1	0.6424	0.2712	0.0931	0.0274	0.0070	0.0016	0.0003	0.0001	0.0000	0.0000
2	0.8729	0.5371	0.2537	0.0982	0.0321	0.0090	0.0021	0.0004	0.0001	0.0000
3	0.9659	0.7636	0.4711	0.2340	0.0962	0.0332	0.0097	0.0024	0.0005	0.0001
4	0.9928	0.9020	0.6821	0.4207	0.2137	0.0905	0.0320	0.0095	0.0023	0.0005
5	0.9988	0.9666	0.8385	0.6167	0.3783	0.1935	0.0826	0.0294	0.0086	0.0020
6	0.9998	0.9905	0.9305	0.7800	0.5611	0.3407	0.1734	0.0736	0.0258	0.0073
7	1.0000	0.9977	0.9745	0.8909	0.7265	0.5118	0.3061	0.1536	0.0639	0.0216
8	1.0000	0.9995	0.9920	0.9532	0.8506	0.6769	0.4668	0.2735	0.1340	0.0539
9	1.0000	0.9999	0.9979	0.9827	0.9287	0.8106	0.6303	0.4246	0.2424	0.1148
10	1.0000	1.0000	0.9995	0.9944	0.9703	0.9022	0.7712	0.5858	0.3843	0.2122
11	1.0000	1.0000	0.9999	0.9985	0.9893	0.9558	0.8746	0.7323	0.5426	0.3450
12	1.0000	1.0000	1.0000	0.9996	0.9966	0.9825	0.9396	0.8462	0.6937	0.5000
13	1.0000	1.0000	1.0000	0.9999	0.9991	0.9940	0.9745	0.9222	0.8173	0.6550
14	1.0000	1.0000	1.0000	1.0000	0.9998	0.9982	0.9907	0.9656	0.9040	0.7878
15	1.0000	1.0000	1.0000	1.0000	1.0000	0.9995	0.9971	0.9868	0.9560	0.8852
16	1.0000	1.0000	1.0000	1.0000	1.0000	0.9999	0.9992	0.9957	0.9826	0.9461
17	1.0000	1.0000	1.0000	1.0000	1.0000	1.0000	0.9998	0.9988	0.9942	0.9784
18	1.0000	1.0000	1.0000	1.0000	1.0000	1.0000	1.0000	0.9997	0.9984	0.9927
19	1.0000	1.0000	1.0000	1.0000	1.0000	1.0000	1.0000	0.9999	0.9996	0.9980
20	1.0000	1.0000	1.0000	1.0000	1.0000	1.0000	1.0000	1.0000	0.9999	0.9995
21	1.0000	1.0000	1.0000	1.0000	1.0000	1.0000	1.0000	1.0000	1.0000	0.9999
22	1.0000	1.0000	1.0000	1.0000	1.0000	1.0000	1.0000	1.0000	1.0000	1.0000

$p=$	0.05	0.10	0.15	0.20	0.25	0.30	0.35	0.40	0.45	0.50
$n=30, x=0$	0.2146	0.0424	0.0076	0.0012	0.0002	0.0000	0.0000	0.0000	0.0000	0.0000
1	0.5535	0.1837	0.0480	0.0105	0.0020	0.0003	0.0000	0.0000	0.0000	0.0000
2	0.8122	0.4114	0.1514	0.0442	0.0106	0.0021	0.0003	0.0000	0.0000	0.0000
3	0.9392	0.6474	0.3217	0.1227	0.0374	0.0093	0.0019	0.0003	0.0000	0.0000
4	0.9844	0.8245	0.5245	0.2552	0.0979	0.0302	0.0075	0.0015	0.0002	0.0000
5	0.9967	0.9268	0.7106	0.4275	0.2026	0.0766	0.0233	0.0057	0.0011	0.0002
6	0.9994	0.9742	0.8474	0.6070	0.3481	0.1595	0.0586	0.0172	0.0040	0.0007
7	0.9999	0.9922	0.9302	0.7608	0.5143	0.2814	0.1238	0.0435	0.0121	0.0026
8	1.0000	0.9980	0.9722	0.8713	0.6736	0.4315	0.2247	0.0940	0.0312	0.0081
9	1.0000	0.9995	0.9903	0.9389	0.8034	0.5888	0.3575	0.1763	0.0694	0.0214
10	1.0000	0.9999	0.9971	0.9744	0.8943	0.7304	0.5078	0.2915	0.1350	0.0494
11	1.0000	1.0000	0.9992	0.9905	0.9493	0.8407	0.6548	0.4311	0.2327	0.1002
12	1.0000	1.0000	0.9998	0.9969	0.9784	0.9155	0.7802	0.5785	0.3592	0.1808
13	1.0000	1.0000	1.0000	0.9991	0.9918	0.9599	0.8737	0.7145	0.5025	0.2923
14	1.0000	1.0000	1.0000	0.9998	0.9973	0.9831	0.9348	0.8246	0.6448	0.4278
15	1.0000	1.0000	1.0000	0.9999	0.9992	0.9936	0.9699	0.9029	0.7691	0.5722
16	1.0000	1.0000	1.0000	1.0000	0.9998	0.9979	0.9876	0.9519	0.8644	0.7077
17	1.0000	1.0000	1.0000	1.0000	0.9999	0.9994	0.9955	0.9788	0.9286	0.8192
18	1.0000	1.0000	1.0000	1.0000	1.0000	0.9998	0.9986	0.9917	0.9666	0.8998
19	1.0000	1.0000	1.0000	1.0000	1.0000	1.0000	0.9996	0.9971	0.9862	0.9506
20	1.0000	1.0000	1.0000	1.0000	1.0000	1.0000	0.9999	0.9991	0.9950	0.9786
21	1.0000	1.0000	1.0000	1.0000	1.0000	1.0000	1.0000	0.9998	0.9984	0.9919
22	1.0000	1.0000	1.0000	1.0000	1.0000	1.0000	1.0000	1.0000	0.9996	0.9974
23	1.0000	1.0000	1.0000	1.0000	1.0000	1.0000	1.0000	1.0000	0.9999	0.9993
24	1.0000	1.0000	1.0000	1.0000	1.0000	1.0000	1.0000	1.0000	1.0000	0.9998
25	1.0000	1.0000	1.0000	1.0000	1.0000	1.0000	1.0000	1.0000	1.0000	1.0000

$p =$	0.05	0.10	0.15	0.20	0.25	0.30	0.35	0.40	0.45	0.50
$n = 40, x = 0$	0.1285	0.0148	0.0015	0.0001	0.0000	0.0000	0.0000	0.0000	0.0000	0.0000
1	0.3991	0.0805	0.0121	0.0015	0.0001	0.0000	0.0000	0.0000	0.0000	0.0000
2	0.6767	0.2228	0.0486	0.0079	0.0010	0.0001	0.0000	0.0000	0.0000	0.0000
3	0.8619	0.4231	0.1302	0.0285	0.0047	0.0006	0.0001	0.0000	0.0000	0.0000
4	0.9520	0.6290	0.2633	0.0759	0.0160	0.0026	0.0003	0.0000	0.0000	0.0000
5	0.9861	0.7937	0.4325	0.1613	0.0433	0.0086	0.0013	0.0001	0.0000	0.0000
6	0.9966	0.9005	0.6067	0.2859	0.0962	0.0238	0.0044	0.0006	0.0001	0.0000
7	0.9993	0.9581	0.7559	0.4371	0.1820	0.0553	0.0124	0.0021	0.0002	0.0000
8	0.9999	0.9845	0.8646	0.5931	0.2998	0.1110	0.0303	0.0061	0.0009	0.0001
9	1.0000	0.9949	0.9328	0.7318	0.4395	0.1959	0.0644	0.0156	0.0027	0.0003
10	1.0000	0.9985	0.9701	0.8392	0.5839	0.3087	0.1215	0.0352	0.0074	0.0011
11	1.0000	0.9996	0.9880	0.9125	0.7151	0.4406	0.2053	0.0709	0.0179	0.0032
12	1.0000	0.9999	0.9957	0.9568	0.8209	0.5772	0.3143	0.1285	0.0386	0.0083
13	1.0000	1.0000	0.9986	0.9806	0.8968	0.7032	0.4408	0.2112	0.0751	0.0192
14	1.0000	1.0000	0.9996	0.9921	0.9456	0.8074	0.5721	0.3174	0.1326	0.0403
15	1.0000	1.0000	0.9999	0.9971	0.9738	0.8849	0.6946	0.4402	0.2142	0.0769
16	1.0000	1.0000	1.0000	0.9990	0.9884	0.9367	0.7978	0.5681	0.3185	0.1341
17	1.0000	1.0000	1.0000	0.9997	0.9953	0.9680	0.8761	0.6885	0.4391	0.2148
18	1.0000	1.0000	1.0000	0.9999	0.9983	0.9852	0.9301	0.7911	0.5651	0.3179
19	1.0000	1.0000	1.0000	1.0000	0.9994	0.9937	0.9637	0.8702	0.6844	0.4373
20	1.0000	1.0000	1.0000	1.0000	0.9998	0.9976	0.9827	0.9256	0.7870	0.5627
21	1.0000	1.0000	1.0000	1.0000	1.0000	0.9991	0.9925	0.9608	0.8669	0.6821
22	1.0000	1.0000	1.0000	1.0000	1.0000	0.9997	0.9970	0.9811	0.9233	0.7852
23	1.0000	1.0000	1.0000	1.0000	1.0000	0.9999	0.9989	0.9917	0.9595	0.8659
24	1.0000	1.0000	1.0000	1.0000	1.0000	1.0000	0.9996	0.9966	0.9804	0.9231
25	1.0000	1.0000	1.0000	1.0000	1.0000	1.0000	0.9999	0.9988	0.9914	0.9597
26	1.0000	1.0000	1.0000	1.0000	1.0000	1.0000	1.0000	0.9996	0.9966	0.9808
27	1.0000	1.0000	1.0000	1.0000	1.0000	1.0000	1.0000	0.9999	0.9988	0.9917
28	1.0000	1.0000	1.0000	1.0000	1.0000	1.0000	1.0000	1.0000	0.9996	0.9968
29	1.0000	1.0000	1.0000	1.0000	1.0000	1.0000	1.0000	1.0000	0.9999	0.9989
30	1.0000	1.0000	1.0000	1.0000	1.0000	1.0000	1.0000	1.0000	1.0000	0.9997
31	1.0000	1.0000	1.0000	1.0000	1.0000	1.0000	1.0000	1.0000	1.0000	0.9999
32	1.0000	1.0000	1.0000	1.0000	1.0000	1.0000	1.0000	1.0000	1.0000	1.0000

$p=$	0.05	0.10	0.15	0.20	0.25	0.30	0.35	0.40	0.45	0.50
$n=50, x=0$	0.0769	0.0052	0.0003	0.0000	0.0000	0.0000	0.0000	0.0000	0.0000	0.0000
1	0.2794	0.0338	0.0029	0.0002	0.0000	0.0000	0.0000	0.0000	0.0000	0.0000
2	0.5405	0.1117	0.0142	0.0013	0.0001	0.0000	0.0000	0.0000	0.0000	0.0000
3	0.7604	0.2503	0.0460	0.0057	0.0005	0.0000	0.0000	0.0000	0.0000	0.0000
4	0.8964	0.4312	0.1121	0.0185	0.0021	0.0002	0.0000	0.0000	0.0000	0.0000
5	0.9622	0.6161	0.2194	0.0480	0.0070	0.0007	0.0001	0.0000	0.0000	0.0000
6	0.9882	0.7702	0.3613	0.1034	0.0194	0.0025	0.0002	0.0000	0.0000	0.0000
7	0.9968	0.8779	0.5188	0.1904	0.0453	0.0073	0.0008	0.0001	0.0000	0.0000
8	0.9992	0.9421	0.6681	0.3073	0.0916	0.0183	0.0025	0.0002	0.0000	0.0000
9	0.9998	0.9755	0.7911	0.4437	0.1637	0.0402	0.0067	0.0008	0.0001	0.0000
10	1.0000	0.9906	0.8801	0.5836	0.2622	0.0789	0.0160	0.0022	0.0002	0.0000
11	1.0000	0.9968	0.9372	0.7107	0.3816	0.1390	0.0342	0.0057	0.0006	0.0000
12	1.0000	0.9990	0.9699	0.8139	0.5110	0.2229	0.0661	0.0133	0.0018	0.0002
13	1.0000	0.9997	0.9868	0.8894	0.6370	0.3279	0.1163	0.0280	0.0045	0.0005
14	1.0000	0.9999	0.9947	0.9393	0.7481	0.4468	0.1878	0.0540	0.0104	0.0013
15	1.0000	1.0000	0.9981	0.9692	0.8369	0.5692	0.2801	0.0955	0.0220	0.0033
16	1.0000	1.0000	0.9993	0.9856	0.9017	0.6839	0.3889	0.1561	0.0427	0.0077
17	1.0000	1.0000	0.9998	0.9937	0.9449	0.7822	0.5060	0.2369	0.0765	0.0164
18	1.0000	1.0000	0.9999	0.9975	0.9713	0.8594	0.6216	0.3356	0.1273	0.0325
19	1.0000	1.0000	1.0000	0.9991	0.9861	0.9152	0.7264	0.4465	0.1974	0.0595
20	1.0000	1.0000	1.0000	0.9997	0.9937	0.9522	0.8139	0.5610	0.2862	0.1013
21	1.0000	1.0000	1.0000	0.9999	0.9974	0.9749	0.8813	0.6701	0.3900	0.1611
22	1.0000	1.0000	1.0000	1.0000	0.9990	0.9877	0.9290	0.7660	0.5019	0.2399
23	1.0000	1.0000	1.0000	1.0000	0.9996	0.9944	0.9604	0.8438	0.6134	0.3359
24	1.0000	1.0000	1.0000	1.0000	0.9999	0.9976	0.9793	0.9022	0.7160	0.4439
25	1.0000	1.0000	1.0000	1.0000	1.0000	0.9991	0.9900	0.9427	0.8034	0.5561
26	1.0000	1.0000	1.0000	1.0000	1.0000	0.9997	0.9955	0.9686	0.8721	0.6641
27	1.0000	1.0000	1.0000	1.0000	1.0000	0.9999	0.9981	0.9840	0.9220	0.7601
28	1.0000	1.0000	1.0000	1.0000	1.0000	1.0000	0.9993	0.9924	0.9556	0.8389
29	1.0000	1.0000	1.0000	1.0000	1.0000	1.0000	0.9997	0.9966	0.9765	0.8987
30	1.0000	1.0000	1.0000	1.0000	1.0000	1.0000	0.9999	0.9986	0.9884	0.9405
31	1.0000	1.0000	1.0000	1.0000	1.0000	1.0000	1.0000	0.9995	0.9947	0.9675
32	1.0000	1.0000	1.0000	1.0000	1.0000	1.0000	1.0000	0.9998	0.9978	0.9836
33	1.0000	1.0000	1.0000	1.0000	1.0000	1.0000	1.0000	0.9999	0.9991	0.9923
34	1.0000	1.0000	1.0000	1.0000	1.0000	1.0000	1.0000	1.0000	0.9997	0.9967
35	1.0000	1.0000	1.0000	1.0000	1.0000	1.0000	1.0000	1.0000	0.9999	0.9987
36	1.0000	1.0000	1.0000	1.0000	1.0000	1.0000	1.0000	1.0000	1.0000	0.9995
37	1.0000	1.0000	1.0000	1.0000	1.0000	1.0000	1.0000	1.0000	1.0000	0.9998
38	1.0000	1.0000	1.0000	1.0000	1.0000	1.0000	1.0000	1.0000	1.0000	1.0000

Percentage points of the Normal distribution

The values z in the table are those which a random variable $Z \sim N(0, 1)$ exceeds with probability p; that is, $P(Z > z) = 1 - \phi(z) = p$

p	z	p	z
0.5000	0.0000	0.0500	1.6449
0.4000	0.2533	0.0250	1.9600
0.3000	0.5244	0.0100	2.3263
0.2000	0.8416	0.0050	2.5758
0.1500	1.0364	0.0010	3.0902
0.1000	1.2816	0.0005	3.2905

Critical values for correlation coefficients

These tables concern tests of the hypothesis that a population correlation coefficient ρ is 0. The values in the tables are the minimum values which need to be reached by a sample correlation coefficient in order to be significant at the level shown, on a one-tailed test.

Product Moment Coefficient level					Sample level	Spearman's Coefficient Level		
0.100	0.050	0.025	0.010	0.005		0.050	0.025	0.010
0.8000	0.9000	0.9500	0.9800	0.9900	4	1.0000	—	—
0.6870	0.8054	0.8783	0.9343	0.9587	5	0.9000	1.0000	1.0000
0.6084	0.7293	0.8114	0.8822	0.9172	6	0.8286	0.8857	0.9429
0.5509	0.6694	0.7545	0.8329	0.8745	7	0.7143	0.7857	0.8929
0.5067	0.6215	0.7067	0.7887	0.8343	8	0.6429	0.7381	0.8333
0.4716	0.5822	0.6664	0.7498	0.7977	9	0.6000	0.7000	0.7833
0.4428	0.5494	0.6319	0.7155	0.7646	10	0.5636	0.6485	0.7455
0.4187	0.5214	0.6021	0.6851	0.7348	11	0.5364	0.6182	0.7091
0.3981	0.4973	0.5760	0.6581	0.7079	12	0.5035	0.5874	0.6783
0.3802	0.4762	0.5529	0.6339	0.6835	13	0.4835	0.5604	0.6484
0.3646	0.4575	0.5324	0.6120	0.6614	14	0.4637	0.5385	0.6264
0.3507	0.4409	0.5140	0.5923	0.6411	15	0.4464	0.5214	0.6036
0.3383	0.4259	0.4973	0.5742	0.6226	16	0.4294	0.5029	0.5824
0.3271	0.4124	0.4821	0.5577	0.6055	17	0.4142	0.4877	0.5662
0.3170	0.4000	0.4683	0.5425	0.5897	18	0.4014	0.4716	0.5501
0.3077	0.3887	0.4555	0.5285	0.5751	19	0.3912	0.4596	0.5351
0.2992	0.3783	0.4438	0.5155	0.5614	20	0.3805	0.4466	0.5218
0.2914	0.3687	0.4329	0.5034	0.5487	21	0.3701	0.4364	0.5091
0.2841	0.3598	0.4227	0.4921	0.5368	22	0.3608	0.4252	0.4975
0.2774	0.3515	0.4132	0.4815	0.5256	23	0.3528	0.4160	0.4862
0.2711	0.3438	0.4044	0.4716	0.5151	24	0.3443	0.4070	0.4757
0.2653	0.3365	0.3961	0.4622	0.5052	25	0.3369	0.3977	0.4662
0.2598	0.3297	0.3882	0.4534	0.4958	26	0.3306	0.3901	0.4571
0.2546	0.3233	0.3809	0.4451	0.4869	27	0.3242	0.3828	0.4487
0.2497	0.3172	0.3739	0.4372	0.4785	28	0.3180	0.3755	0.4401
0.2451	0.3115	0.3673	0.4297	0.4705	29	0.3118	0.3685	0.4325
0.2407	0.3061	0.3610	0.4226	0.4629	30	0.3063	0.3624	0.4251
0.2070	0.2638	0.3120	0.3665	0.4026	40	0.2640	0.3128	0.3681
0.1843	0.2353	0.2787	0.3281	0.3610	50	0.2353	0.2791	0.3293
0.1678	0.2144	0.2542	0.2997	0.3301	60	0.2144	0.2545	0.3005
0.1550	0.1982	0.2352	0.2776	0.3060	70	0.1982	0.2354	0.2782
0.1448	0.1852	0.2199	0.2597	0.2864	80	0.1852	0.2201	0.2602
0.1364	0.1745	0.2072	0.2449	0.2702	90	0.1745	0.2074	0.2453
0.1292	0.1654	0.1966	0.2324	0.2565	100	0.1654	0.1967	0.2327

Random numbers

86 13	84 10	07 30	39 05	97 96	88 07	37 26	04 89	13 48	19 20
60 78	48 12	99 47	09 46	91 33	17 21	03 94	79 00	08 50	40 16
78 48	06 37	82 26	01 06	64 65	94 41	17 26	74 66	61 93	24 97
80 56	90 79	66 94	18 40	97 79	93 20	41 51	25 04	20 71	76 04
99 09	39 25	66 31	70 56	30 15	52 17	87 55	31 11	10 68	98 23
56 32	32 72	91 65	97 36	56 61	12 79	95 17	57 16	53 58	96 36
66 02	49 93	97 44	99 15	56 86	80 57	11 78	40 23	58 40	86 14
31 77	53 94	05 93	56 14	71 23	60 46	05 33	23 72	93 10	81 23
98 79	72 43	14 76	54 77	66 29	84 09	88 56	75 86	41 67	04 42
50 97	92 15	10 01	57 01	87 33	73 17	70 18	40 21	24 20	66 62
90 51	94 50	12 48	88 95	09 34	09 30	22 27	25 56	40 76	01 59
31 99	52 24	13 43	27 88	11 39	41 65	00 84	13 06	31 79	74 97
22 96	23 34	46 12	67 11	48 06	99 24	14 83	78 37	65 73	39 47
06 84	55 41	27 06	74 59	14 29	20 14	45 75	31 16	05 41	22 96
08 64	89 30	25 25	71 35	33 31	04 56	12 67	03 74	07 16	49 32
86 87	62 43	15 11	76 49	79 13	78 80	93 89	09 57	07 14	40 74
94 44	97 13	77 04	35 02	12 76	60 91	93 40	81 06	85 85	72 84
63 25	55 14	66 47	99 90	02 90	83 43	16 01	19 69	11 78	87 84
11 22	83 98	15 21	18 57	53 42	91 91	26 52	89 13	86 00	47 61
01 70	10 83	94 71	13 67	11 12	36 54	53 32	90 43	79 01	95 15

Objectives checklist

Ch	Objective	MyMaths	InvisiPen	No	Almost	Yes!
12 Algebra 2	Make logical deductions and prove statements directly by exhaustion, by counter example and by contradiction.	2254	12S1A	☐	☐	☐
	Understand and use functions.	2049, 2135, 2138, 2139, 2142, 2261	12S2B	☐	☐	☐
	Understand and use parametric equations.	2224, 2262	12S3A	☐	☐	☐
	Understand and use algebraic fractions in all their forms.	2200, 2259	12S4B	☐	☐	☐
	Decompose fractions into partial fractions.	2260	12S5A	☐	☐	☐
13 Sequences	Use the binomial expansion and recognise the range of validity.	2204, 2205	13S1A	☐	☐	☐
	Use the binomial expansion to estimate the value of a surd.	2204, 2205	–	☐	☐	☐
	Understand if a sequence is increasing or decreasing.	2264	13S2B	☐	☐	☐
	Work out the order of a periodic sequence.	2264	–	☐	☐	☐
	Work out the nth term and the sum of an arithmetic series.	2039	13S3A	☐	☐	☐
	Work out the nth term and the sum of a geometric series.	2040	13S4B	☐	☐	☐
	Evaluate a series given in sigma notation.	2040	–	☐	☐	☐
14 Trigonometric identities	Convert between degrees and radians and use radians in problems.	2050, 2266	14S1A	☐	☐	☐
	Use reciprocal and inverse trigonometric functions.	2155, 2156	14S2B	☐	☐	☐
	Use trigonometric formulae for compound angles, double angles and half angles.	2157, 2158, 2262	14S3A	☐	☐	☐
	Find and use equivalent forms for $a\cos\theta + b\sin\theta$	2159	14S4B	☐	☐	☐
	Solve equations using trigonometric formulae to simplify expressions.	2159	14S4B	☐	☐	☐
15 Differentiation 2	Find points of inflection and determine when a curve is convex or concave.	2271	15S1A	☐	☐	☐
	Use $\lim\limits_{\theta\to0}\dfrac{\sin\theta}{\theta}=1$ and $\lim\limits_{\theta\to0}\dfrac{1-\cos\theta}{\theta}=0$	2165	–	☐	☐	☐
	Differentiate $\sin x$, $\cos x$, e^x, a^x and $\ln x$	2165, 2161	15S2B, 15S3A	☐	☐	☐
	Use the product and quotient rule for differentiation.	2163, 2164	15S4B	☐	☐	☐
	Use the chain rule for differentiation.	2162, 2166	15S5A	☐	☐	☐
	Find the derivative of a function that is defined implicitly.	2223	15S7A	☐	☐	☐
	Find the derivative of an inverse function.	2272	15S6B	☐	☐	☐
	Find the derivative of a function that is defined parenthetically.	2222	15S8B	☐	☐	☐

Objectives checklist

Ch	Objective	MyMaths	InvisiPen	No	Almost	Yes!
16 Integration and differential equations	Integrate a set of standard functions, f(x) and the related functions, f(ax + b)	2057, 2168, 2170, 2218	16SA	☐	☐	☐
	Find the area between two curves.	2274	–	☐	☐	☐
	Simplify an integral by changing the variable, referred to as *substitution*.	2167, 2169, 2216, 2219	16S2B	☐	☐	☐
	Use integration by parts to integrate the product of two functions.	2171, 2220	16S3A	☐	☐	☐
	Simplify an integral by decomposing a rational function into partial fractions.	2217	16S4B	☐	☐	☐
	Understand the meaning of the expression 'differential equation'.	2226, 2227	16S5A	☐	☐	☐
	Use integration where the variables are separable.	2226, 2227	16S5A	☐	☐	☐
17 Numerical Methods	Use the change of sign method to find and estimate the root(s) of an equation.	2173	17S1B	☐	☐	☐
	Use an iterative formula to estimate the root of an equation.	2174	17S2A	☐	☐	☐
	Recognise the conditions that cause an iterative sequence to converge.	2174	–	☐	☐	☐
	Use the Newton-Raphson method to estimate the root of an equation.	2176	17S3B	☐	☐	☐
	Use the trapezium rule to find the area under a curve.	2060	17S4A	☐	☐	☐
18 Motion in two dimensions	Use the constant acceleration equations for motion in two dimensions.	2290	18S1B	☐	☐	☐
	Use calculus to solve problems in two-dimensional motion with variable acceleration.	2291	18S2A	☐	☐	☐
	Solve problems involving the motion of a projectile under gravity.	2198, 2199	18S3B	☐	☐	☐
	Analyse the motion of an object in two dimensions under the action of a system of forces.	2192	18S4A	☐	☐	☐
19 Forces 2	Manipulate vectors in three dimensions, and solve geometrical problems.	2208	–	☐	☐	☐
	Understand that there is a maximum value that the frictional force can take (μR) and that it takes this value when the object is moving or on the point of moving.	2193	–	☐	☐	☐
	Resolve in suitable directions to find unknown forces when the system is at rest or has constant acceleration.	2190, 2191	19S2A	☐	☐	☐
	Use constant acceleration formulae for problems involving blocks on slopes or blocks connected by pulleys.	2191	19S1B	☐	☐	☐
	Solve differential equations which arise from problems involving $F = ma$	2194	–	☐	☐	☐
	Take moments about suitable points and resolve in suitable directions to find unknown forces.	2197	19S3B	☐	☐	☐

Ch	Objective	MyMaths	InvisiPen	No	Almost	Yes!
20 Probability and continuous random variables	Calculate conditional probabilities from data given in different forms.	2092, 2095	20S1A	☐	☐	☐
	Apply binomial and Normal probability models in different circumstances.	2113, 2120, 2121, 2292	20S2B, 20S3A	☐	☐	☐
	Use data to assess the validity of probability models.	2286	—	☐	☐	☐
	Solve problems involving both binomial and Normal distributions.	2286	20S4B	☐	☐	☐
21 Hypothesis testing 2	State null and alternative hypotheses when testing for correlation.	2287	21S1A	☐	☐	☐
	Compare a given PMCC to a critical value or its p-value to the significance level, and use this comparison to decide whether to accept or reject the null hypothesis.	2287	21S1A	☐	☐	☐
	Decide what the conclusion means in context about the correlation.	2287	21S1A	☐	☐	☐
	State null and alternative hypotheses when testing the mean of a Normal distribution.	2288	21S2B	☐	☐	☐
	Calculate the test statistic, compare it to a critical value or compare its p-value to the significance level, and use this comparison to decide whether to accept or reject the null hypothesis.	2288	21S2B	☐	☐	☐
	Decide what the conclusion means in context about the mean of the distribution.	2288	21S2B	☐	☐	☐

Answers

Paper 1 (Set A)

1 $x\ln 2x - x + c$

2 a 6 cm   **b** $(6\pi - 9\sqrt{3})\ \text{cm}^2$

3 a i Gradient of 1st line $= 2$ Gradient of 2nd line $= 2$
       Gradient of 3rd line $= -2$ Gradient of 4th line $= -2$

     ii $ABCD$ is a parallelogram because it has two pairs of parallel sides.

   b i Solve simultaneously each pair of non-parallel equations.

     ii $\left(-\dfrac{1}{8}, \dfrac{3}{4}\right), \left(\dfrac{5}{4}, \dfrac{7}{2}\right), \left(\dfrac{21}{8}, -\dfrac{19}{4}\right), (4, -2)$

   c 3.07 cm (to 3 sf), 6.15 cm (to 3 sf)

4 a Assume that the contrary is true i.e. $\sqrt{2}$ is rational.
     Then there exist integers p and q (with no common factors)
     such that $\sqrt{2} = \dfrac{p}{q}$

     $\Rightarrow 2 = \dfrac{p^2}{q^2} \Rightarrow p^2 = 2q^2 \Rightarrow p^2$ is even $\Rightarrow p$ is even.

     If p is even, then it is a multiple of $2 \Rightarrow p = 2n$, where n is an integer.

     Also, $\sqrt{2} = \dfrac{p}{q} \Rightarrow q^2 = \dfrac{p^2}{2} \Rightarrow q^2 = \dfrac{(2n)^2}{2} = 2n^2 \Rightarrow q^2$ is even $\Rightarrow q$ is even.

     But if p and q are both even, then they have a common factor of 2
     This **contradicts** our original assumption.
     $\sqrt{2}$ is **not** rational.
     $\therefore \sqrt{2}$ is irrational.

   b $\sqrt[8]{2}$

   c i $4\sqrt{2}$ **ii** $16 + 8\sqrt{2}$

5 a $1 < k < 4$

   b i 6   **ii** $3x + 4$

6 a $x = -7, x = 2$

   b $y = \dfrac{1}{2}x + 15$

   c $\dfrac{119}{4} + 14\ln 2$

7 a $\dfrac{\cos^2\theta}{\cos^2\theta} + \dfrac{\sin^2\theta}{\cos^2\theta} \equiv \dfrac{1}{\cos^2\theta}$

     $\Rightarrow 1 + \tan^2\theta \equiv \sec^2\theta$

   b $\theta = -0.905,\ \theta = 0.905,\ \theta = 5.38$ (to 3 sf)

   c $x = -2.26,\ x = 2.26,\ x = 4.02$ (to 3 sf)

8 a Let $f(x) = x^2 - 6$
     Then $f(2.4) = -0.24 < 0$ and $f(2.5) = 0.25 > 0$
     $f(x)$ is **continuous** on the interval $(0.24, 0.25)$ and there is a **change of sign**.
     $\therefore f(x) = 0$ has a root between $x = 2.4$ and $x = 2.5$

   b 2.45

   c 1.320469 (to 6 dp)

   d No, her procedure will not succeed as the gradient of the curve at x_0 is zero (or close to zero).

9 a $\sec^2 y$ **b** $\dfrac{1}{1 + x^2}$

Paper 2 (Set A)

1 $\ln 5$

2 Stretch parallel to the x-axis with scale factor 2
    Stretch parallel to the y-axis with scale factor 8

---

3 $\sqrt{\dfrac{11}{3}},\ (-1, 2)$

4 $\dfrac{3}{4x} + \dfrac{2}{x^2} - \dfrac{7}{4(x-4)} + \dfrac{1}{x-5}$

5 a i

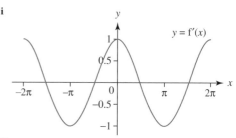

     ii $y = \cos x$

   b i

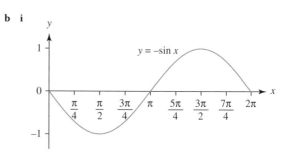

     ii $y = -\sin x$

6 a π

   b Both radius (or radius squared) and area cannot be negative.

   c i $y = \sqrt{r^2 - x^2}$

       Area required $= \displaystyle\int_0^r \sqrt{r^2 - x^2}\ dx$

       Using $x = r\sin\theta$,

       $x = 0 \Rightarrow \theta = \sin^{-1}\left(\dfrac{0}{r}\right) = 0$

       $x = r \Rightarrow \theta = \sin^{-1}\left(\dfrac{r}{r}\right) = \dfrac{\pi}{2}$

       Then area required becomes

       $\displaystyle\int_0^{\frac{\pi}{2}} \sqrt{r^2 - r^2\sin^2\theta}\ \dfrac{dx}{d\theta}\ d\theta$

       $= r\displaystyle\int_0^{\frac{\pi}{2}} \sqrt{1 - \sin^2\theta}\ (r\cos\theta)d\theta$

       $= r^2\displaystyle\int_0^{\frac{\pi}{2}} \cos^2\theta\ d\theta$

     ii $\dfrac{1}{4}\pi r^2$

   d πr^2

7 a -1260 **b** $\dfrac{1}{8}$ **c** $\dfrac{3925}{16}$

8 18.5 N, 240.3°

9 a

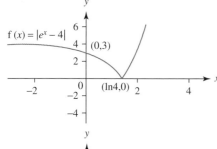

$f(x) = |e^x - 4|$ (0,3)   (ln4,0)

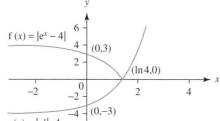

$f(x) = |e^x - 4|$ (0,3)   (ln 4,0)   (0,−3)

$g(x) = |e^x| - 4$

b

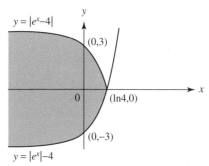

$y = |e^x - 4|$ (0,3)   (ln4,0)   (0,−3)

$y = |e^x| - 4$

10 a   Distance between $(0, 0)$ and (x, y) is

$$D = \sqrt{(y-0)^2 + (x-0)^2}$$
$$= \sqrt{x^2 + y^2}$$
$$y = 4x + 3 \Rightarrow D = \sqrt{x^2 + (4x+3)^2}$$
$$\therefore D = \sqrt{17x^2 + 24x + 9}$$

b   **i** $\left(-\dfrac{12}{17}, \dfrac{3}{17}\right)$ **ii** $\dfrac{3\sqrt{17}}{17}$

11 a   **i**   Domain of $f(x)$ $-1 \le x \le 1$ Range of $f(x)$ $-\dfrac{\pi}{2} \le y \le \dfrac{\pi}{2}$

        Domain of $g(x)$ $-1 \le x \le 1$ Range of $g(x)$ $0 \le y \le \pi$

    ii

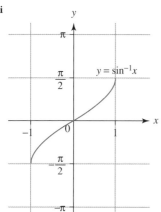

$y = \sin^{-1} x$

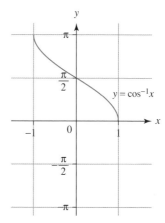

$y = \cos^{-1} x$

b $\dfrac{\pi}{2}$

c   Let $\theta = \sin^{-1} x$

    Then $x = \sin\theta$

    So $\cos(2\sin^{-1} x) = \cos 2\theta$

                  $= 1 - 2\sin^2\theta$

                  $= 1 - 2x^2$

12 a $2, (1, 2)$

    b   **i** $y = 4$ **ii** $2\sqrt{21}$ **iii**   20.8 (to 3 sf)

Paper 3 (Set A)

1   **a**   0.159   **b**   0.066 8

2

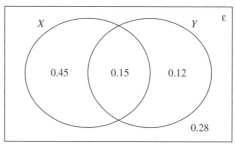

X Y ε

0.45   0.15   0.12   0.28

3   No. You would expect roughly $270 \div 6 = 45$ of each outcome, but 72 is very much more than expected.

4   **a** $X \sim B(37\,890, 0.3)$

    b np is very large.

       p is not too far from 0.5

       $Y \sim N(37\,890 \times 0.3, 37\,890 \times 0.3 \times 0.7)$

       $= N(11\,367, 7956.9)$

    c $P(11\,000 \le X \le 11\,500)$

       $\approx P(10\,999.5 \le Y \le 11\,500.5)$

       $= P(Y \le 11\,500.5) - P(Y \le 10\,999.5)$

       $= P\left(Z \le \dfrac{11\,500.5 - 11\,367}{\sqrt{7956.9}}\right) - P\left(Z \le \dfrac{10\,999.5 - 11\,367}{\sqrt{7956.9}}\right)$

       $= 0.932\,75 - 0.000\,018\,9 \approx 0.932\,7$

5   **a, b, c, d**

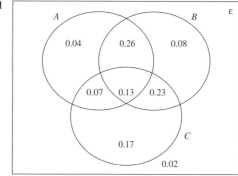

A B ε

0.04   0.26   0.08

0.07   0.13   0.23

0.17 C 0.02

6 a Continuous. The data has been rounded for presentation but time takes continuous values.

b 5.21, 4.97

c 5.37, 5.27

d

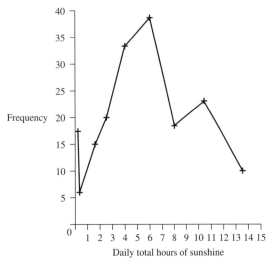

Daily total hours of sunshine

7 a i 2   **ii** 2.55

b i Upper quartile $+ 1.5 \times$ IQR *or* lower quartile $- 1.5 \times$ IQR

ii There is one outlier – Cambourne.

c Value at (2, 4.5) circled.

d i Line through two calculated points.

ii weak positive

iii approx 7 knots

iv It is just a prediction; the relationship is not exact.

e i $H_0: \rho = 0$ $H_1 : \rho \neq 0$ where ρ is the population correlation coefficient.

ii $0.520 < 0.754$

The result is not significant at the 5% level.

There is insufficient evidence to suggest that there is any correlation between levels of rainfall and windspeed.

8 196 N

9 30.9 N m (to 3 sf) clockwise

10 a i gradient of graph = acceleration of the particle

ii $a = \dfrac{\Delta v}{\Delta t}$

$a = \dfrac{v - u}{t}$

$at = v - u$

$v = u + at$

b 20 m (to nearest metre)

11 a Perpendicular to the wall

Since their wall is modelled as smooth, there is no friction to cause a component of the reaction in the vertical direction.

b 35.7 N

12 2.13 m s^{-2}

13 a 41.3°   **b** 7.8 s

14 a

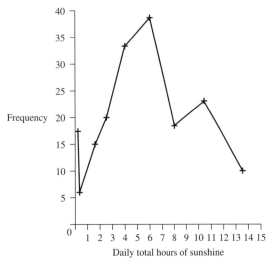

b 37.5 N

c 4.05 m s^{-2}

15 a $\dfrac{20\pi}{6}\cos\left(\dfrac{\pi}{6}t\right)\mathbf{i} - 3t^{-\frac{1}{2}}\mathbf{j}$

b i $\dfrac{100\pi}{6}\cos\left(\dfrac{\pi}{6}t\right)\mathbf{i} - 15t^{-\frac{1}{2}}\mathbf{j}$

ii $t = 9 \Rightarrow \mathbf{F} = \dfrac{100\pi}{6}\cos\left(\dfrac{9\pi}{6}\right)\mathbf{i} - 15 \times 9^{-\frac{1}{2}}\mathbf{j}$

$= 0\mathbf{i} - 5\mathbf{j}$

Hence $|\mathbf{F}| = 5$

Force acts in the negative **j**-direction.

c $\mathbf{r} = \left(100 - \dfrac{120}{\pi}\cos\left(\dfrac{\pi}{6}t\right)\right)\mathbf{i} + \left(208 - 4t^{\frac{3}{2}}\right)\mathbf{j}$

Paper 1 (Set B)

1 $x < y, \; x \leq -y$

2 $\dfrac{\mathrm{d}y}{\mathrm{d}x} = 16(3 - 5x)^2(4x - 3)^3 - 10(4x - 3)^4(3 - 5x)$

$= 2(3 - 5x)(4x - 3)^3[8(3 - 5x) - 5(4x - 3)]$

$= 2(3 - 5x)(4x - 3)^3(39 - 60x)$

$= 6(3 - 5x)(13 - 20x)(4x - 3)^3$

3 a

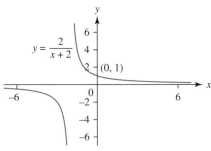

b

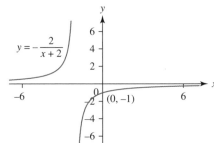

c

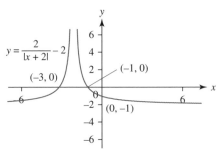

4 a Method 1: Substitution

$x = t\left(\dfrac{1}{1 + t^2}\right) = ty$

$\Rightarrow t = \dfrac{x}{y}$

Substituting into $y = \dfrac{1}{t^2 + 1}$,

$y = \dfrac{1}{\left(\dfrac{x}{y}\right)^2 + 1}$

$= \dfrac{y^2}{x^2 + y^2}$

$\Rightarrow x^2 + y^2 - y = 0$

Method 2: Combining x and y directly

$$x^2 + y^2 = \frac{1+t^2}{(1+t^2)^2}$$
$$= \frac{1}{1+t^2}$$
$$= y$$
$$\Rightarrow x^2 + y^2 - y = 0$$

b $1 + (x-3)^2 = \dfrac{y^2}{16}$

5 a $\dfrac{\mathrm{d}T}{\mathrm{d}t}$ is the rate of change of temperature.

k is the constant of proportionality.
k is negative because the temperature is decreasing.
$(T - T_0)$ is the difference between the temperature of the body and the temperature of the surroundings.

b i $\dfrac{1}{10}\ln 2$ **ii** $35\,^\circ\mathrm{C}$ **iii** 25 minutes and 51 seconds

c T will converge to $20\,^\circ\mathrm{C}$
The tea will cool to the temperature of its surroundings.

6 a $27^2 = 729$, $9^3 = 729$

b $50 = 3 + 21 + 36$ (or equivalent eg $1 + 21 + 28$ etc.)
$51 = 6 + 45$ (or equivalent)
$52 = 1 + 6 + 45$ (or equivalent)
$53 = 10 + 15 + 28$ (or equivalent)
$54 = 3 + 6 + 45$ (or equivalent)
$55 = 55$ (or equivalent)
$56 = 1 + 55$ (or equivalent)
$57 = 1 + 1 + 55$ (or equivalent)
$58 = 3 + 55$ (or equivalent)
$59 = 1 + 3 + 55$ (or equivalent)
$60 = 15 + 45$ (or equivalent)
Fiona's claim is correct.

c Assume the contrary, i.e. the number of primes is finite.
Let the primes be called $P_1, P_2, P_3, \ldots, P_n$
Let a new number, Q be equal to the product of **all** of the primes, plus 1
i.e. $Q = (P_1\, P_2\, P_3\, \ldots\, P_n) + 1$

There are only two possibilities:
1. Q is prime and since $Q \neq P_1, P_2, P_3, \ldots, P_n$; this contradicts our original assumption that the only primes were $P_1, P_2, P_3, \ldots, P_n$
2. Q is not prime.
By the principle of prime factorisation, Q must be divisible by a prime number $P_1, P_2, P_3, \ldots, P_n$

However, dividing Q by any of $P_1, P_2, P_3, \ldots, P_n$ leaves a remainder of 1
So Q must be divisible by a different prime, not contained in $P_1, P_2, P_3, \ldots, P_n$
This contradicts our original assumption that the only primes were $P_1, P_2, P_3, \ldots, P_n$

Both arguments result in a contradiction.
$\Rightarrow$ The number of primes is infinite.

7 a i $x \ln x - x + c$

ii Let $u = \ln x \Rightarrow \dfrac{\mathrm{d}u}{\mathrm{d}x} = \dfrac{1}{x}$

$\dfrac{\mathrm{d}v}{\mathrm{d}x} = \ln x \Rightarrow v = x \ln x - x$

$$\int (\ln x)^2 \,\mathrm{d}x = (x \ln x - x)\ln x - \int (\ln x - 1)\,\mathrm{d}x$$
$$= (x \ln x - x)\ln x - (x \ln x - x - x) + d$$
$$= x(\ln x)^2 - 2x \ln x + 2x + d$$
$$= x(\ln x)^2 + 2x(1 - \ln x) + d$$

b $3 - \mathrm{e}$

8 a $2\sin\left(\theta + \dfrac{\pi}{3}\right)$ **b** $\dfrac{-23\pi}{12}, \dfrac{-11\pi}{12}, \dfrac{\pi}{12}, \dfrac{13\pi}{12}$

9 a $2 - \dfrac{x}{192} - \dfrac{5x^2}{147\,456}$ **b** $|x| < 64$ **c** $1.994\,8$

10 a $\overrightarrow{AB} = (5\mathbf{i} + 2\mathbf{j} - \mathbf{k}) - (\mathbf{i} - \mathbf{j} + \mathbf{k})$
$= 4\mathbf{i} + 3\mathbf{j} - 2\mathbf{k}$
$\overrightarrow{DC} = (7\mathbf{i} + 5\mathbf{j} - 3\mathbf{k}) - (3\mathbf{i} + 2\mathbf{j} - \mathbf{k})$
$= 4\mathbf{i} + 3\mathbf{j} - 2\mathbf{k}$
$\overrightarrow{AD} = (3\mathbf{i} + 2\mathbf{j} - \mathbf{k}) - (\mathbf{i} - \mathbf{j} + \mathbf{k})$
$= 2\mathbf{i} + 3\mathbf{j} - 2\mathbf{k}$
$\overrightarrow{BC} = (7\mathbf{i} + 5\mathbf{j} - 3\mathbf{k}) - (5\mathbf{i} + 2\mathbf{j} - \mathbf{k})$
$= 2\mathbf{i} + 3\mathbf{j} - 2\mathbf{k}$
$\overrightarrow{AB} = \overrightarrow{DC}$ and $\overrightarrow{AD} = \overrightarrow{BC}$
$\therefore ABCD$ is a parallelogram.

b $\left|\overrightarrow{AB}\right| = \sqrt{29}$ $\left|\overrightarrow{BC}\right| = \sqrt{17}$

c i $-2\mu\mathbf{i}$ **ii** $\lambda(-5\mathbf{i} - \dfrac{9}{2}\mathbf{j} + 3\mathbf{k})$

11 a $\mathrm{fg}(x) = x^{\frac{1}{12}}$ **b** $\dfrac{2x-1}{x}$ **c** $\mathrm{j}(x) = x^2 - 5x$

Paper 2 (Set B)

1 4

2 $\ln 3^{2x-1} = \ln 20$
$(2x - 1)\ln 3 = \ln 20$
$2x \ln 3 = \ln 20 + \ln 3$
$2x \ln 3 = \ln 60$
$x = \dfrac{\ln 60}{2\ln 3}$
$\Rightarrow x = \dfrac{\ln 60}{\ln 9}$

3 8

4 $p = q - 1$

5 $y + 1 = 2(x - 1)$

6 a Translation with vector $\begin{pmatrix} 4 \\ 3 \end{pmatrix}$

b i domain $x \geq 4$ range $\mathrm{f}(x) \geq 3$
ii

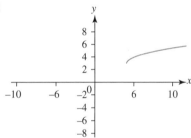

c i domain $x \geq 3$ range $\mathrm{f}^{-1}(x) \geq 4$
ii

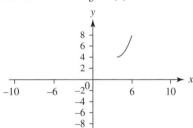

7 a $-2 < x < 2$

b Since $\mathrm{f}''(x)$ changes sign as it passes through $x = 0$, there is a point of inflection at $x = 0$
$\mathrm{f}(0) = 3 \Rightarrow$ inflection point at $(0, 3)$

8 $\mathrm{f}'(\theta) = \lim_{h \to 0} \dfrac{\mathrm{f}(\theta + h) - \mathrm{f}(\theta)}{h}$

$\mathrm{f}'(\theta) = \lim_{h \to 0} \dfrac{\cos(\theta + h) - \cos\theta}{h}$

$\mathrm{f}'(\theta) = \lim_{h \to 0} \dfrac{\cos\theta\cos h - \sin\theta\sin h - \cos\theta}{h}$

$\mathrm{f}'(\theta) = \lim_{h \to 0} \left[\dfrac{(\cos h - 1)}{h}\cos\theta - \dfrac{\sin h}{h}\sin\theta \right]$

As $h \to 0$, $\dfrac{\cos h - 1}{h} \approx \dfrac{(1 - h^2) - 1}{h} = -h \to 0$

and $\dfrac{\sin h}{h} \approx \dfrac{h}{h} \to 1$

$\therefore \mathrm{f}'(x) = -\sin\theta$

9 $\ln\left(\dfrac{C(x-1)}{x}\right)$

10 a $C = \dfrac{5}{9}F - \dfrac{160}{9}$

  b p = Temperature change in Celsius for every $1°$ temperature change in Fahrenheit

  q = Temperature in Celsius when temperature in Fahrenheit is zero

  c $160°$

11 a $x > \dfrac{20}{19}$

  b $x \le -\dfrac{3}{4}, x \ge \dfrac{5}{2}$

  c $-2 < x < -1, \ 1 < x < 2$

12 a

x	2	3	5	6	8
y	7.68	6.144	3.93216	3.145728	2.01326592
$\log_{10}x$	0.301029...	0.477121...	0.698970...	0.778151...	0.903089...
$\log_{10}y$	0.885361...	0.788451...	0.594631...	0.497721...	0.303901...

  b i $\log_{10}y$

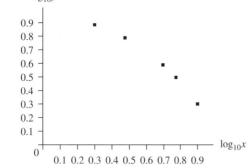

  ii $\log_{10}y$

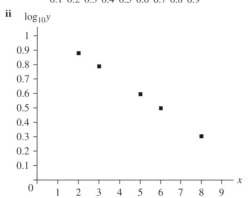

  c i $y = ab^x$

  $\Rightarrow \log_{10} y = \log_{10} a + x \log_{10} b$

  $\Rightarrow$ Graph of $\log_{10} y$ against x gives a straight line, so data obeys this relationship.

  ii $12, 0.8$

13 Left hand side gives

$$\frac{\sin^2 2x}{4} + \sin^4 x$$

$$= \frac{(2\sin x \cos x)^2}{4} + \sin^4 x$$

$$= \sin^2 x \cos^2 x + \sin^4 x$$

$$= \sin^2 x(\cos^2 x + \sin^2 x)$$

$$= \sin^2 x$$

Right hand side gives

$$1 - \frac{\cot^2 x}{\operatorname{cosec}^2 x}$$

$$= 1 - \frac{1}{\operatorname{cosec}^2 x}(\operatorname{cosec}^2 x - 1)$$

$$= 1 - \sin^2 x\left(\frac{1}{\sin^2 x} - 1\right)$$

$$= 1 - (1 - \sin^2 x)$$

$$= \sin^2 x$$

$$= \text{Left hand side}$$

14 a $y = 6(t + 2)(t - 2) = 0$ for two different values of t i.e. $t = 2$ and $t = -2$

For both values, $x = 0$

$\therefore$ Curve crosses itself at $(0,0)$

  b $y = 3x$ and $y = -3x$

Paper 3 (Set B)

1 0.6

2 a 0.4207 **b** 0.6554 **c** 0.2347

3 a i 0.3 **ii** 0.564 **iii** 0.226

  b $P(\text{green} \mid \text{plain}) = \dfrac{50}{27 + 23 + 50} = 0.5$

  $P(\text{green}) = \dfrac{75 + 50}{250} = 0.5$

  These are equal, so the events are independent.

4 a $P(X = 0) = P(X = 4) = \dfrac{k}{9}$ $\quad P(X = 1) = P(X = 3) = \dfrac{k}{3}$

  $P(X = 2) = k$

  $k = \dfrac{9}{17}$

  b i Y follows a continuous distribution, so a continuity correction must be used.

  ii $P(X = 0 \text{ or } 4) = 0.02232$

  $P(X = 1 \text{ or } 3) = 0.2297$

  $P(X = 2) = 0.4950$

  c It reflects the right shape of the distribution in terms of being symmetric about 2, but the values are too different.

5 a Stratified – When the population can be divided into separate groups which you expect to be represented differently for the survey, you sample from each group in proportion to its size in the population.

Systematic – Find a sample of size n from a population of size N by taking one member at random from the first k members of the population and then every k^{th} member after that, where $k = \dfrac{N}{n}$

  b The weather stations and the records of the various measurements held by the stations.

  c Decide how many weather stations you want to take data from that day and find the sampling fraction.

Take an ordered list of the weather stations (perhaps by how far north they lie) and take the first random station and all the other stations from the list at regular intervals afterwards.

6 a i, iii

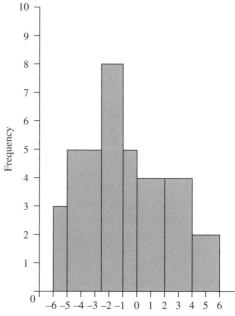

Difference between temperatures on consecutive days

  ii -6 to -5 occurs with frequency 3

  -5 to -2.5 occurs with frequency 5

6 b The graph is symmetric about a clear mean value.
However, there are more outcomes than you might expect in the –6 to –5 range.

c –0.955 deg C (to 3 sf), 8.51 (to 3 sf)

d 33.7 deg C (to 3 sf), 27.6 (to 3 sf)

e $H_0 : \mu = 0$ $H_1 : \mu \neq 0$

where μ is the mean difference in temperature.

Either

The test statistic is $\dfrac{-0.954\,8 - 0}{\sqrt{\dfrac{8.508\,9}{31}}} = -1.822\,5$

The critical value is $-1.959\,9$

$-1.822\,5 > -1.959\,9$ so the result is not significant.

or

The test statistic has a p-value of 3.42%.

3.42% > 2.5% so the result is not significant.

Do not reject H_0 because there is insufficient evidence to suggest that there is a difference between corresponding days' temperatures on average.

7 $11\mathbf{i} - 2\mathbf{j}$

8 20.8 N

9 $\mathbf{T} = 49.6$ N, $\mathbf{R} = 65.3$ N

10 a 8.6 m (to 2 sf)   **b** 17.1 m s^{-1} at 30.1° below horizontal

11 a

$5.8v$ N

696 N or 71g N

b 120 m s^{-1}

c $\mathbf{F} = m\mathbf{a}$

$$696 - 5.8v = 71\dfrac{\mathrm{d}v}{\mathrm{d}t}$$

$$5.8(120 - v) = 71\dfrac{\mathrm{d}v}{\mathrm{d}t}$$

$$120 - v = 12.2\dfrac{\mathrm{d}v}{\mathrm{d}t}$$

d 16.9 s (to 3 sf)

12 a 1.43   **b** 5.35 N, 0.89 m s^{-1}

Paper 1 (Set C)

1 $x + \ln 2$

2 $\dfrac{3 + x}{3 + 3x}$

3 a The integrand does not exist (or is not defined) when $x = 0$

The area would be infinite, as $\dfrac{1}{x^2} \to \infty$ as $x \to 0$

$\dfrac{1}{x^2}$ not continuous on $[-2, 3]$, and the Fundamental Theorem of Calculus only applies to continuous functions.

Sally is correct.

b $\dfrac{1}{x^2}$ is always positive, so the area underneath its graph should be positive.

However, Mark's calculation gives a negative answer so must be incorrect.

4 a i 37.651

ii

Overestimate.

Curve is concave up, which means that the curve lies completely below the top of each strip. Thus the area under the curve is less than the sum of the areas under all of the strips.

b i $(x + 3)^2 = x^2 + 6x + 9$

Also, $x(x - 1) + a(x - 1) + b$

$$= x^2 - x + ax - a + b$$

$$= x^2 + (a - 1)x + (b - a)$$

For $(x + 3)^2 = x(x - 1) + a(x - 1) + b$, we need:

$a - 1 = 6 \Rightarrow a = 7$

$b - a = 9 \Rightarrow b = 16$

ii $20 + 16\ln 3$

iii 0.2%

5 a $-\dfrac{8}{9}$ **b** 1   **c** $\dfrac{1}{2}$

6 a $x^2 + y^2 + 8x - 6y - 11 = 0$

$\Rightarrow (x + 4)^2 - 16 + (y - 3)^2 - 9 - 11 = 0$

$\Rightarrow (x + 4)^2 + (y - 3)^2 = 36$

$\Rightarrow$ Centre $(-4, 3)$ and radius 6

$x^2 + y^2 - 16x - 16y + 79 = 0$

$\Rightarrow (x - 8)^2 - 64 + (y - 8)^2 - 64 + 79 = 0$

$\Rightarrow (x - 8)^2 + (y - 4)^2 = 49$

$\Rightarrow$ Centre $(8, 8)$ and radius 7

Distance between centres

$= \sqrt{(8 - 3)^2 + (8 - -4)^2}$

$= 13$

Sum of the two radii $= 6 + 7 = 13$

$=$ Distance between centres

$\therefore$ Circles touch.

b i Point of contact is $\dfrac{6}{13}$ of the way along the line segment from $(-4, 3)$ to $(8, 8)$ ie $\left[-4 + \dfrac{6}{13}(8 - (-4)), \; 3 + \dfrac{6}{13}(8 - 3) \right]$

$= \left(\dfrac{20}{13}, \dfrac{69}{13} \right)$

ii Tangent at $\left(\dfrac{20}{13}, \dfrac{69}{13} \right)$ will be perpendicular to the line joining the two centres (the normal to both circles).

$y - \dfrac{69}{13} = -\dfrac{12}{5}\left(x - \dfrac{20}{13} \right)$

7 a $\dfrac{\mathrm{d}P}{\mathrm{d}t} = kP$

b $\displaystyle\int \dfrac{\mathrm{d}P}{P} = \int k \; \mathrm{d}t$

$\Rightarrow \ln P = kt + \ln A$

$\Rightarrow P = \mathrm{e}^{kt}\mathrm{e}^{\ln A}$

$P = A\mathrm{e}^{kt}$

c 5.2×10^6, $\dfrac{1}{14}\ln\left(\dfrac{15}{13} \right)$, 8.15×10^6 (to 3 sf)

8 $4(x + 2)^2 + 25(y + 1)^2 = 100$

9 a Letting $u = \cos x$ gives

$\dfrac{\mathrm{d}u}{\mathrm{d}x} = -\sin x \Rightarrow \mathrm{d}x = -\dfrac{\mathrm{d}u}{\sin x}$

Hence

$\displaystyle\int \tan x \; \mathrm{d}x = \int \dfrac{\sin x}{\cos x}\mathrm{d}x$

$\displaystyle = \int -\dfrac{\sin x}{u} \dfrac{\mathrm{d}u}{\sin x}$

$\displaystyle = \int -\dfrac{1}{u}\mathrm{d}u$

$= -\ln|u| + c$

$= -\ln|\cos x| + c$

$= \ln\left|(\cos x)^{-1}\right| + c$

$= \ln|\sec x| + c$

9 b Method 1: Letting $u = \sec x$

Then $\dfrac{du}{dx} = \sec x \tan x \Rightarrow dx = \dfrac{du}{\sec x \tan x}$

$\displaystyle \int \sec^4 x \tan x \, dx = \int \sec^3 x \left(\sec x \tan x \right) \, dx$

$\displaystyle = \int u^3 (\sec x \tan x) \, \frac{du}{(\sec x \tan x)}$

$\displaystyle = \int u^3 \, du$

$\displaystyle = \frac{u^4}{4} + c$

$\displaystyle = \frac{\sec^4 x}{4} + c$

OR Method 2: Letting $u = \tan x$

$\dfrac{du}{dx} = \sec^2 x \Rightarrow dx = \dfrac{du}{\sec^2 x}$

$\displaystyle \int \sec^4 x \tan x \, dx$

$\displaystyle = \int \sec^2 x \tan x (\sec^2 x) \, dx$

$\displaystyle = \int (1 + \tan^2 x) \tan x (\sec^2 x) \, dx$

$\displaystyle = \int (1 + u^2) u (\sec^2 x) \frac{du}{(\sec^2 x)}$

$\displaystyle = \int (1 + u^2) u \, du$

$\displaystyle = \int u + u^3 \, du$

$\displaystyle = \frac{u^2}{2} + \frac{u^4}{4} + k$

$\displaystyle = \frac{\tan^2 x}{2} + \frac{\tan^4 x}{4} + k$

$\displaystyle = \frac{\sec^2 x - 1}{2} + \frac{(\sec^2 x - 1)^2}{4} + k$

$\displaystyle = \frac{2\sec^2 x - 2}{4} + \frac{\sec^4 x - 2\sec^2 x + 1}{4} + k$

$\displaystyle = \frac{\sec^4 x - 1}{4} + k$

$\displaystyle = \frac{\sec^4 x}{4} + C, \text{ where } C = k - \frac{1}{4}$

c The following uses method 2 in part **b**.

$\displaystyle \int [\sec x \tan x (\sec^3 x) - \sec x + \tan x] \, dx$

$\displaystyle = \int \sec^4 x \tan x \, dx - \int \sec x \, dx + \int \tan x \, dx$

$\displaystyle = \frac{\sec^4 x - 1}{4} - \ln|\sec x + \tan x| + \ln|\sec x| + \ln A$ (where $\ln A$ is an arbitrary constant)

$\displaystyle = \frac{\sec^4 x - 1}{4} + \ln\left(\frac{A \sec x}{\sec x + \tan x} \right)$

10 a $20x^4 - 72x^3 + 87x^2 - 38x + 3$
$= (10x^2 - 11x + 1)(ax^2 + bx + c)$

Equating coefficients of x^4:
$20 = 10a \Rightarrow a = 2$
Equating constants:
$3 = c$
Equating coefficients of x^3:
$-72 = 10b - 11a \Rightarrow b = -5$
$\therefore 20x^4 - 72x^3 + 87x^2 - 38x + 3$
$= (10x^2 - 11x + 1)(2x^2 - 5x + 3)$
$\Rightarrow (10x^2 - 11x + 1)$ is a quadratic factor

b $(10x - 1)(2x - 3)(x - 1)^2$

c

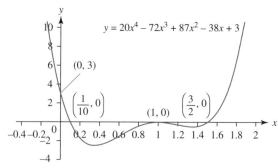

d Reflection in the y-axis.

11 a i $\dfrac{3}{n^2}\left[\dfrac{n(n+1)}{2} \right] + 4 = \dfrac{3(n+1)}{2n} + 4$

$= \dfrac{3}{2} + \dfrac{3}{2n} + 4$

$= \dfrac{11}{2} + \dfrac{3}{2n}$

ii $S_n = \dfrac{n}{2}\left[2(1) + (n-1)(1) \right]$

$S_n = \dfrac{n}{2}[2 + n - 1]$

$S_n = \dfrac{n(n+1)}{2}$

b i $c_i = 0 + \left(\dfrac{1-0}{n} \right) i \Rightarrow c_i = \dfrac{i}{n}$

ii $\dfrac{11}{2}$

iii $\dfrac{11}{2}$

Paper 2 (Set C)

1 $|x| < \dfrac{b}{a}$

2 $x = Ae^{kt}$

3 a $\dfrac{1}{4} x^{-2} e^{\frac{1}{2}x} (x - 2)$ **b** $\ln|\ln x| + c$

4 $12 - 4\sqrt{3}$ units, $10 - 4\sqrt{3}$ units

5 $2x^2 - 12x + 19 = 2(x^2 - 6x) + 19$
$= 2\left[(x - 3)^2 - 9 \right] + 19$
$= 2(x - 3)^2 + 1 \ (x - 3)^2 \geq 0 \Rightarrow 2(x - 3)^2 + 1 > 0$

6 a Let $f(x) = x^3 - x^2 + 3x - 4$
$f(1) = -1 < 0$
$f(1.5) = 1.625 > 0$
$f(x)$ is **continuous** and there is a **change of sign**.
$\Rightarrow 1 < \alpha < 1.5$

b $x^3 = x^2 + 4 - 3x$
$x^2 = \dfrac{x^2 + 4}{x} - 3$
$x = \sqrt{\dfrac{x^2 + 4}{x} - 3}$

c 1.22

d

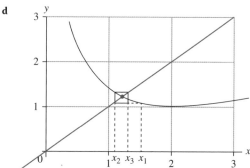

7 $\dfrac{2(xy^2 - x - \sin^2 y)}{4(x + \sin^2 y)\sin y \cos y - 2x^2 y}$

8 a i $\dfrac{2\pi}{b}$ **ii** 1

   b i $\dfrac{2\pi}{b}$ **ii** a

   c i $\dfrac{2\pi}{b}$ **ii** a

   d $y = 4.75\cos\dfrac{\pi}{6}x + 12.25$

9 $-\cosec 2y$

10 Yes, as his solution is equivalent to $-a < x < a$

11 a

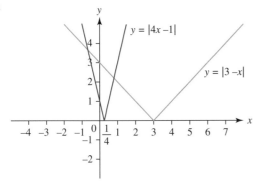

   b $\left(-\dfrac{2}{3}, \dfrac{11}{3}\right), \left(\dfrac{4}{5}, \dfrac{11}{5}\right)$

   c $-\dfrac{2}{3} < x < \dfrac{4}{5}$

12 $\dfrac{3}{8}$

13 Let the two integers be A and B
Then the two-digit numbers are
AB and BA
Adding, we have
$10A + B + 10B + A$
$= 11A + 11B$
$= 11(A + B)$
which is a multiple of 11
$\therefore$ Vince's observation is always true.

14 Translation with vector $\begin{pmatrix} -\dfrac{\pi}{12} \\ 0 \end{pmatrix}$

15 a $2, \sqrt{2}, 1$ **b** $\dfrac{1}{8} + \dfrac{1}{8}\sqrt{2}$

16 a $\sin\left(\dfrac{\pi}{2} - A - B\right) \equiv \sin\left(\dfrac{\pi}{2} - A\right)\cos B - \cos\left(\dfrac{\pi}{2} - A\right)\sin B$

     $\equiv \cos A \cos B - \sin A \sin B$

     But $\sin\left(\dfrac{\pi}{2} - A - B\right)$

     $= \sin\left[\dfrac{\pi}{2} - (A + B)\right] = \cos(A + B)$

     $\therefore \cos(A + B) \equiv \cos A \cos B - \sin A \sin B$

   b Replacing B by A in **a**, we have
     $\cos(A + A) \equiv \cos A \cos A - \sin A \sin A$
     $\cos 2A \equiv \cos^2 A - \sin^2 A$
     Using $\sin^2 A \equiv 1 - \cos^2 A$, we get
     $\cos 2A = \cos^2 A - (1 - \cos^2 A)$
     $\Rightarrow \cos 2A = 2\cos^2 A - 1$
     $\Rightarrow \cos^2 A = \dfrac{1 + \cos 2A}{2}$

   c $\dfrac{1}{32}(12A + 8\sin 2A + \sin 4A) + c$

Paper 3 (Set C)

1 Independent means knowing that one event will occur does not affect the probability that the other event will occur.

2 a (very) weak positive

   b $H_0 : \rho = 0$ $H_1 : \rho \neq 0$
     Where ρ is the population correlation coefficient between the scores on the two tests.
     Since $0.268\,4 < 0.514$ there is insufficient evidence to reject H_0.
     You conclude that there is no correlation between scores on the two tests.

3 a The events are mutually exclusive and their probabilities sum to 1

   b

EEEE	EEEO	EEOE	EEOO
EOEE	EOEO	EOOE	EOOO
OEEE	OEEO	OEOE	OEOO
OOEE	OOEO	OOOE	OOOO

   c $\dfrac{5}{16}$

   d $\dfrac{5}{11}$

4 a Only those who are available in the middle of the day will take part, e.g. the unemployed. This could bias the sample.

   b $X \sim B\left(30, \dfrac{1}{3}\right)$

   c $10, \dfrac{20}{3}$

   d 0.035 5 (to 3 sf)

   e $H_0 : p = \dfrac{1}{3}$ $H_1 : p > \dfrac{1}{3}$

     where p is the probability that a randomly chosen person supports the policy.
     Either
     Since $P(X \geq 14) = 8.98\%$
     and $P(X \geq 15) = 4.35\%$,
     Or
     The critical region is $X \geq 15$
     The result (15 people) is contained within the critical region.
     Half the people is 15 supporting the policy.
     $P(X \geq 15) = 4.35\% < 5\%$
     Reject H_0
     There is sufficient evidence at the 5% level to suggest that the company's assumption was an under-estimate.

5 a 23 knots   **b** UQ = 26, LQ = 17, IQR = 9   **c** $\dfrac{57}{388}$

   d

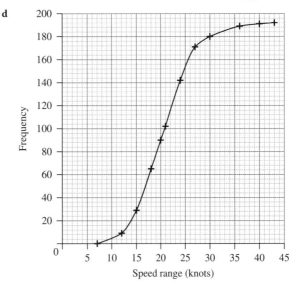

   e The central tendencies are very similar, as are the spreads.

6 a

Hurn
0 2 4 6 8 10 12 14 16

   b Largest: Camborne
     Smallest: Leuchars

   c No outlier, IQR = 6.5
     $UQ + 1.5 \times 6.5 = 16.25$
     $LQ - 1.5 \times 6.5 = -9.25$

   d Heathrow has the most similar 5-number summary values to the others on average.

   e Advantage: It is easier and cheaper to take a sample.
     Disadvantage: Could succumb to bias from regional variations.

   f A simple random sample is one in which every possible sample of the desired size is equally likely to be chosen.
     Taking a sample in this way definitely excludes all entries not from the chosen location.

7 area between graph and x-axis

8 a 53.1°

   b $\dfrac{3}{5}\mathbf{i} - \dfrac{4}{5}\mathbf{j}$

9 a 140 N   **b** 1.6 m

10 $-\dfrac{3}{4}\mathbf{i} + \mathbf{j}\ \text{m s}^{-2}$

11 a $v = \dfrac{ds}{dt} = \dfrac{d}{dt}(te^{5-t})$
$$= e^{5-t} - te^{5-t}$$
$$= -e^{5-t}(t-1)\ \text{ms}^{-1}$$

   b $6e^4$ N

12 2.75 m (to 3 sf)

13 a $x = ut\cos\theta$
$$y = ut\sin\theta - \frac{1}{2}gt^2$$
$$t = \frac{x}{u\cos\theta}$$
$$y = u\frac{x}{u\cos\theta}\sin\theta - \frac{1}{2}g\left(\frac{x}{u\cos\theta}\right)^2$$
$$= x\tan\theta - \frac{1}{2}g\frac{x^2}{u^2\cos^2\theta}$$
$$y = x\tan\theta - \frac{g\sec^2\theta}{2u^2}x^2$$

   b 74.3°, 29.7°

   c The minimum angle would be greater. The maximum angle would be smaller.
     Air resistance would mean that the horizontal range of the jump would be less for each value of θ